A FEELING FOR THE ORGANISM

Barbara McClintock's graduation portrait, Cornell University, 1923.

A FEELING FOR THE ORGANISM

The Life and Work of Barbara McClintock

Evelyn Fox Keller

Massachusetts Institute of Technology

A W. H. FREEMAN / HOLT PAPERBACK

Henry Holt and Company New York

Holt Paperbacks
Henry Holt and Company, LLC
Publishers since 1866
175 Fifth Avenue
New York, New York 10010
www.henryholt.com

A Holt Paperback® and ® are registered trademarks
of Henry Holt and Company, LLC.

Library of Congress Cataloging-in-Publication Data
Keller, Evelyn Fox, 1936–
A feeling for the organism.
p. cm.
Includes index.
ISBN-13: 978-0-8050-7458-1
ISBN-10: 0-8050-7458-9
1. McClintock, Barbara, 1902–. 2. Geneticists—United States—
Biography. I. Title.
QH429.2.M38K44 1983
575.1'092'4 [B] 82-21066

Henry Holt books are available for special promotions and
premiums. For details contact: Director, Special Markets.

Originally published in hardcover in 1983
by W. H. Freeman and Company

First Holt Paperbacks Edition 2003

A W. H. Freeman / Holt Paperback

Designed by Gary A. Head

Printed in the United States of America
10 9 8

To Sarah and Jeffrey

Barbara McClintock receiving the 1983 Nobel Prize for physiology or medicine from King Carl Gustaf of Sweden. (UPI/Bettmann)

Author's Note on the Tenth Anniversary Edition

Five months after this book first appeared, Barbara McClintock was awarded the Nobel Prize. With that award, she was instantly transformed into a scientific heroine and a darling of the media. Inevitably, she was ambivalent about the attention—the incursion on her privacy, not to mention the romanticization—but she had to be gratified by the recognition finally accorded to her extraordinary work. Certainly, her friends and admirers were elated—as, of course, was I.

Since 1983, McClintock's place in the history of modern biology has only become more secure. Transposition is now a standard and crucial element of an emerging new biology of development and evolution. Biology and the times have moved on, and new histories will inevitably be written. Nonetheless, the story I told still seems to me essentially just, and I have decided to let this book stand as I initially wrote it.

Contents

Foreword

Barbara McClintock was the kind of person whom it is good to meet, stimulating to converse with, and rewarding to understand. But for years those privileges were enjoyed by only a small circle of scientists, the conversation was pretty technical, and the understanding slow to come. Evelyn Fox Keller's insightful biography can now interpret and extend these pleasures to a broad crescent of readers who care about interesting people or about how science advances. In particular, I hope it will bring scientists-to-be a deeper understanding of one of the exemplary paths to scientific achievement. And—for a culture that believes itself to be an enlightened, thoughtful one—Keller's calm recital of how McClintock faced professional gender hurdles and prejudices is factual reportage that can give every reader, male or female, a vicarious experience of these problems. Moreover, her analysis of McClintock's scientific work—in its broad context—describes some difficult aspects of modern genetics and itself constitutes a significant contribution to the broad history of thought.

McClintock was gifted and innovative in the use of her hands and of her mind. Born in Connecticut, she devoted her life to one of the quintessential plants of the Americas—corn, or maize—working as an expert geneticist, and when that work demanded it, laboring as a farmer who developed a keen eye for the weather. She brought direct demonstrations of some of the most important links between structure and function that are fundamental in genetics. Chromosomes (structures) had already been recognized by Sturtevant and Bridges of the Morgan school to be the carriers of gene functions, by astute use of circumstantial evidence. McClintock showed soon afterward how we could recognize, before our eyes, individual chromosomes and segments of chromosomes unmistakably undergoing the precise exchanges previously inferred—and then passing the very traits they carried on to the progeny plants. Her several increasingly intricate studies (beginning about 1931) became favorite teaching demonstrations to prove the fundamental "rules" of the material basis of genetics. All students knew of her early work.

But then, she detected some irregularities that seemed to break the rules—partly "reshuffling the genetic cards"—and proceeded alone to work out the higher order of new rules that finally made these irregularities predictable too. How these novel findings were slowly, almost reluctantly, accepted into the logic of her field, being crowned thirty years later with the award of the 1983 Nobel Prize (in McClintock's eighty-second year), is a particularly well-told part of the story in this book. It reminds us that the much praised "simplest hypothesis"—no matter how hypnotic and charming—often requires patching or modification, and thus may be only a first step in the path to understanding.

One quickly noted Barbara McClintock's vitality. Her conversation was clipped, rapid, and purposive—her eyes were bright, observant, and intense. She was quick to pick up the gist of a scientific paper and had a tremendous grasp and knowledge of details from the genetic literature. Her originality made her a free spirit, following stern disciplines of her own rather than conforming to imposed ones. She avoided marriage and family life—even at times advised talented familiars of its traps. Though something of a loner, she could be warm in one-to-one contacts and personable enough to accommodate to many social occasions; she dressed in a unisexual mode from an early time. As a humanist, rather than a feminist, she expected unprejudiced respect for herself and other women. Always wanting to communicate, she strove in later years to control her almost explosive style of speech—tempering it easily to the *level* of her listeners, but much less successfully to their *thinking rate or attention span.*

After her ideas were more widely accepted, and the Nobel award came, many scientists took the opportunity to recall their personal association with Barbara McClintock and even claimed an "early recognition" of her work. I do not think these are entirely expressions of vanity or exhibitionism—there is much to suggest that a lot of the outpouring has also to do with sharing a genuine appreciation of her character and accomplishments—even with a desire to "set the record straight."

My own conversations with her began probably from about 1950. As a chemist, likewise interested in the material basis of genetic action but with a shaky grip on cytogenetics, I remember asking her for an operational definition of just what distinguished heterochromatin from other chromosome substance. Her answer soon turned into a torrent of information about meiotic and developmental processes in corn, which sounded authoritative but left me far behind. After many such sessions, I eventually grasped her fundamental postulates about Dissociator and Activator; without sufficient "feel for the organism" of field corn, I didn't develop my own conviction beyond my admiration for hers. Like many others, I felt bewilderment rather than doubt about the masses of information she reported.

I was mostly on the receiving end of those conversations, but I believe that she had something like the same passive awareness and perhaps respect for my growing evidence that what was truly DNA actually functioned in a fully genetic manner in bacteria. Molecular units did not play much of a part in my story then, because through most of the 1950s we hadn't a clear understanding of the sizes and limits of DNA molecules, but she was never much concerned with chemistry as mechanism. In that sense she lagged somewhat behind a few advanced geneticists, though she had achieved masterful control of the dyes and stains with which she identified individual chromosomes and of the biochemical tools that revealed some of their actions. Later on we talked a good deal about the fates of the Cold Spring Harbor Laboratory, of the Carnegie Division, and of science itself. Keller's descriptions include excerpts that portray tellingly the spirit and style of such exchanges.

I think I understand the nature of the supposed "delay" in acceptance of the later McClintock work. An expert microscopist first of all, she had studied innumerable minute details scrutinized at the high power setting, always making the all-important step of returning to survey in the lower power to see just where she was "on the big scale." But she also had an inner "viewing lens" that mentally portrayed a far bigger map—a panorama of the whole cellular and anatomical display—spreading over into the fourth, or time, dimension of what happened to the chromosomes in various cells and tissues during the whole development of the mature corn plant. With such a vivid inner vision, and a rapid-fire delivery, she would often leap the full range from microscopic observation—to model—to conclusion—to result in the grown plant, within a sentence, or a brief "cluster of words." And back and forth! Characteristically, her descriptions remained verbal; I do not recall her ever stopping to draw for us a sketch or diagram of what she was talking about. Small wonder that most of us were ill prepared—and, I think, too lazy—to work hard enough to master the data as it poured forth. Certainly she felt unfulfilled by the responses most colleagues could give, and she even became diffident about publication. I think most of us admired her greatly, and what resistance there was came from those already lulled into hypnosis by their own countless happy descriptions to students of the classic minuet of the chromosomes—including the contributions McClintock had made earlier. And it is easy to see why she perceived their relative ignorance as negative and came perhaps to dramatize it a bit as rejection or scorn.

Luckily, Evelyn Fox Keller has created here her own word pictures that summarize the adventures of meandering genes. With, of course, the help of professional advances made in the meantime and hindsight, she has done the still hard work of personally mastering these ideas and rendering them more accessible to all of us. She has not stopped with a romantic or intuitive grasp of them—nor does she trivialize or treat scornfully the honest misunderstandings or difficulties that came up. She works to convey the pitfalls and obstacles the actual participants traversed and moreover to see them in their own time frame. Because this story involves the consolidation of cytogenetics, gene action, and molecular biology the enlightenment she achieves is real and lasting.

There are many different valuable lessons to be drawn from this presentation of Barbara McClintock's career. In addition to those already mentioned, several chapters well illustrate the role of hypothesis and model-building in biology. Keller provides a good analysis of the use of language by scientists—the steps by which it becomes more, rather than less, precise—the logical (and personal!) gaps that can be subjective, irrational, subtle, or nonlinear. As perceptive analyst, she notes some of the semantic blocks over which the insiders either leaped or stumbled. Elsewhere she not only tells us what did happen in genetics but also gives us a taste of alternative ideas that in the end were not embraced.

Evelyn Fox Keller has deftly exposed the many scientific, philosophical, personal, social, and historical threads interwoven in this fascinating scientific development. She has bypassed without so much as suggesting, the devices of "science popularization." To have presented sensitively so much excitement and information, embedded in its own prehistory, is in my opinion a splendid achievement that will reward a wide range of readers.

Dr. Rollin D. Hotchkiss
Professor Emeritus, Rockefeller University
Research Professor, Biology, SUNY-Albany

Preface

This book is the biography partly of an individual, but primarily of a relationship. It is the story of the interaction between an individual scientist, Barbara McClintock, known to her colleagues as a maverick and a visionary, and a science, genetics, that has been distinguished in recent decades by extraordinary growth and dramatic transformation. The person and the field are contemporaries, born in the first years of this century. They grew to first maturity in sympathy and mutual prosperity: as a young scientist, McClintock achieved a level of recognition that few women could then imagine. But later years took them along distinct and divergent routes, and McClintock retreated into obscurity. Today, after three decades of estrangement, their paths have begun to converge, and McClintock's name has returned to the center of current biological interest.

Barbara McClintock's involvement with genetics (and cytology) began early enough for her to take part in, and indeed help create, what was then a new area of study. Her accomplishments during her twenties and thirties were summarized by her colleague Marcus Rhoades when she was chosen for the

Kimber Award from the National Academy of Sciences. He wrote:

> One of the remarkable things about Barbara McClintock's surpassingly beautiful investigations is that they came solely from her own labors. Without technical help of any kind, she has by virtue of her boundless energy, her complete devotion to science, her originality and ingenuity, and her quick and high intelligence made a series of significant discoveries unparalleled in the history of cytogenetics.[1]

Later, in her mid-forties, her research in the cytogenetics of maize brought her to concepts so new and so radical that her colleagues had difficulty "hearing" them; and as time progressed, communication became more rather than less difficult. A different revolution—that of molecular biology—had taken center stage. Its success was so dramatic that, by the end of the 1960s, it seemed as if few basic questions were left to be answered. Jacques Monod, one of the central heroes of this drama, could say: "The Secret of Life? But this is in large part known—in principle, if not in all details."[2] Little place seemed to be left for the complexities of McClintock's cytogenetics.

The growing interest in McClintock's work today is a consequence of startling new developments in biology that echo many of the findings she described as long as thirty years ago. In particular, the "transposition" of genetic elements, long regarded as McClintock's invention, has now come to be regarded as an indisputable phenomenon.

Until the late 1960s, genes were thought of as simple units, laid out in a fixed, linear sequence that held the key to the unfolding of the organism. It seemed unreasonable to suppose that genetic elements could spontaneously move from one site to another, even from one chromosome to another, and less reasonable yet to suppose that, as McClintock asserted, such rearrangements might play a crucial role in genetic organization and control. But McClintock has argued ever since the early 1950s that, in moving from one chromosomal site to another, these genetic elements carry new instructions to the cell; furthermore, their very movement is itself "programmed."

New evidence of transposition (now at the molecular level) began to emerge in the late 1960s. Over the past decade, more and more instances of mobile, or transposable, genetic elements have been discovered, suggesting a degree of fluidity of the chromosomal complement (or genome) that is in major conflict with the conceptual framework that has been dominant in biology for at least the past twenty-five years. With the discovery of genetic mobility, our very concept of a genetic program has changed. It needs now to be thought of as a dynamic structure, rather than as a static linear message inscribed in the sequence of DNA. And to the extent that the movement of genes is itself part of the program, we need to ask: Where do the instructions come from? McClintock's answer—that they come from the entire cell, the organism, perhaps even from the environment—is profoundly disturbing to orthodox genetics. For years now, biologists have felt certain that the spectre of adaptive evolution had been laid to rest. But in McClintock's work, the suggestion of environmental influence on genetics surfaces yet again.

To those who see in these recent developments the makings of a new revolution in biology, McClintock's name has become something of a password. Matthew Meselson of Harvard believes that history will "record her as the originator of new and very much more subtle and complex genetic theories that are as yet only dimly understood."[3]

But at the end of her corridor in the lab that bears her name at Cold Spring Harbor, McClintock is still a remote figure. A critical gap remains between her understanding of transposition and that of her colleagues. The rediscovery of transposition was not based on her findings, and thus far has proceeded without her. As she sees it, even today few of her supporters really understand what she has to say. Despite the rapprochement, and the publicity, Barbara McClintock remains in crucial respects an outsider.

It might be tempting to read this history as a tale of dedication rewarded after years of neglect—of prejudice or indifference eventually routed by courage and truth. But the actual story is both more complex and more illuminating. It is a story about the nature of scientific knowledge, and of the tangled web of individual and group dynamics that define its growth.

A new idea, a new conception, is born in the privacy of one man's or one woman's dreams. But for that conception to become part of the body of scientific theory, it must be acknowledged by the society of which the individual is a member. In turn, the collective effort provides the ground out of which new ideas grow. Scientific knowledge as a whole grows out of the interaction—sometimes complex, always subtle—between individual creativity and communal validation. But sometimes that interaction miscarries, and an estrangement occurs between individual and community. Usually, in such a case, the scientist loses credibility. But should that not happen, or, even better, should it happen and then be reversed, we have a special opportunity to understand the meaning of dissent in science.

The story of Barbara McClintock allows us to explore the conditions under which dissent in science arises, the function it serves, and the plurality of values and goals it reflects. It makes us ask: What role do interests, individual and collective, play in the evolution of scientific knowledge? Do all scientists seek the same kinds of explanations? Are the kinds of questions they ask the same? Do differences in methodology between different subdisciplines even permit the same kinds of answers? And when significant differences do arise in questions asked, explanations sought, methodologies employed, how do they affect communication between scientists? In short, why could McClintock's discovery of transposition not be absorbed by her contemporaries? We can say that her vision of biological organization was too remote from the kinds of explanations her colleagues were seeking, but we need to understand what that distance is composed of, and how such divergences develop.

Thomas S. Kuhn has reminded us that conversions in science (or resistances to conversion) occur "not despite the fact that scientists are human but because they are."[4] He chooses to focus attention in the community and the dynamics by which the community forms and reforms itself. Our focus, by contrast, will be on the individual, on the "idiosyncracies of autobiography and personality"[5] that incline an individual scientist to a particular set of methodological and philosophical commitments, to resisting or accepting the dominant trend within a field—*but always against the background of the community.* Of necessity, therefore, this book must serve simultaneously as biography and as intellectual history. Its starting point is the recognition that science is at once a highly personal and a communal endeavor.

An account of the individual's "style" as a scientist—a style partly learned and partly self-generated—is a crucial aspect of this attempt. In the world of contemporary biological research, McClintock's style is highly idiosyncratic. Her passion is for the individual, for difference. "The important thing is to develop the capacity to see one kernel that is different, and make that understandable," she has said. "If [something] doesn't fit, there's a reason, and you find out what it is." McClintock believes that the prevailing focus on classes and numbers encourages researchers to overlook difference, to "call it an exception, an aberration, a contaminant." She sees the consequences as very costly. "Right and left," she says, they miss "what is going on."[6]

A century earlier, McClintock might have been classified among the naturalists. But she would not have quite fit that description either. In a way that has drawn as much from herself as from her milieu, she has succeeded in synthesizing the uniquely twentieth-century focus on experiment with the naturalist's emphasis on observation. The role of vision in her experimental work provides the key to her understanding. What for others is interpretation, or speculation, for her is a matter of trained and direct perception. McClintock has pushed her special blend of observational and cognitive skills so far that few can follow her. She herself cannot quite say how she "knows" what she knows. She talks about the limits of verbally explicit reasoning; she stresses the importance of her "feeling for the organism" in terms that sound like those of mysticism. But like all good mystics, she insists on the utmost critical rigor, and, like all good scientists, her understanding emerges from a thorough absorption in, even identification with, her material.

By her own choice, Barbara McClintock is a recluse. She does not welcome the flood of attention she is now getting; she prefers her own privacy and the autonomy that goes with it. When I first proposed my project to her, she said that she was too much of a maverick for her story to be of interest to others. But I argued that it was illuminating for just that reason. And as the interviews progressed, I drew confirmation for my instinct from her own emphasis on how much can be learned from the exceptional, "anomalous" example.

Gradually I came to understand that the importance of her story lies precisely in her independence. Because of the simplicity with which she pursued what was "obvious" to her, almost independent of response from her peers, I began to see that the person and her work—what goes into her style of investigation and what has come out of it—illustrate each other with special clarity. Because of her often peripheral relation to the community in which she worked, her story provides a view that would normally not be possible into the powerful, shifting currents of group interest.

After my interviews with McClintock, I talked with a number of her colleagues; their remarks help place McClintock's own recollections in perspective. Some of these colleagues have worked with her as classical geneticists, some have esteemed her from afar, and others have only recently come to recognize her voice as important. But I quickly found that, in order to understand the story I was hearing, I had to learn about a science that, as a student of molecular biology in the early 1960s, I had never needed to study. In the book that has emerged, I retrace for the reader, as far as possible, the steps of my own learning process.

I begin in Chapter 1 with a brief sketch of the historical context into which Barbara McClintock is born as a scientist. In Chapter 2, McClintock's own voice is introduced as she recalls the formative years of her childhood. Throughout the remaining chapters, in which her career unfolds to the present day, Barbara McClintock's voice continues to be heard, but increasingly it is interwoven with the words and writings of her colleagues.

By the end, I hope to have drawn a portrait that will function on three levels. For those readers who are not biologists, it will serve as an introduction to an unfamiliar world. To those who have studied classical genetics, it will introduce the person behind the name that is familiar from textbook descriptions of the major landmarks of corn genetics. And professional biologists, working at the forefront of contemporary research, may read this as a book about language—about worlds of discourse that operate to shape the growth of specific areas of research by demarcating them from others. Above all, *A Feeling for the Organism* is the story of one woman's conception of science that gradually—although not irrevocably—isolated her from the evolving discourse of mainstream research. In writing this story, I have been trying to understand the relations between creativity and validation, between individual and community, and between one community's conception of science and another's, that underlie the drama not only of McClintock's story but of scientific research in general.

Evelyn Fox Keller
Cambridge, Massachusetts
October 1982

Acknowledgements

First and foremost, I am indebted to Barbara McClintock for providing the inspiration for this book. The story of her life, which she generously shared with me over the course of a series of interviews, helped to enlarge my understanding as a scientist and as a critic of the dynamics of my own profession. In particular, it gave me a new perspective on the history of molecular biology. I wrote this book, in part, in an attempt to redress certain imbalances, made vivid by McClintock's experiences, that prevail in more conventional accounts of recent biological history. Other works in the social studies of science have also begun to do this; if I have not made explicit reference to such studies, it was because of my belief that the specifics of McClintock's life could best speak for themselves by themselves. My exploration of McClintock's life has also sharpened my thinking on a subject I have written about elsewhere: the relation of gender to science. In her adamant rejection of female stereotypes, McClintock poses a challenge to any simple notions of a "feminine" science. Her pursuit of a life in which "the matter of gender drops away" provides us instead with a glimpse of what a "gender-free" science might look like.

I am additionally grateful to Barbara McClintock for tolerating—even while trying her best to dissuade me from—the writing of this book. She is, of course, in no way responsible for my interpretations. Thanks are also due to her niece, Marjorie Bhavnani, and her sister, Marjorie McKinley, for their patient cooperation and assistance.

My second largest debt is to my editor. No one could wish for a more sympathetic and sensitive editor than Jehane Burns, whose help with this manuscript deserves more thanks than a few words can indicate. Her commitment to and confidence in the meaning I wished to convey repeatedly pushed me to articulate, and even think, my own thoughts more clearly.

I can also only begin to thank the many others who provided me with invaluable assistance in the research for and writing of this book. Harriet Creighton, Marcus Rhoades, George Beadle, Matthew Meselson, David Botstein, Paul and Jean Margolin, Evelyn Witkin, Lotte Auerbach, Richard Lewontin, Tracy Sonneborn, Bruce Wallace, Ruth Sager, and Jim Haber were all more than generous in granting me interviews. Their recollections and impressions provided me with information and insights I could not have gotten elsewhere.

Parts, and in some cases, all, of the manuscript were read by Barbara Ankeny, Carolyn Cohen, Allan Ellis, Stanley Goldberg, Mel Green, Diana Hall, Gerald Holton, Ruth Hubbard, Ken Keniston, Martin Krieger, Tom Kuhn, Ken Manning, Leo Marx, Janet Murray, Peter Peterson, John Richards, Bill and Sally Ruddick, Sam Schweber, Bonnie Sedlak, and Anne Faustoe Sterling. I am grateful to all of them for their many helpful suggestions and their tactful advice. Bonnie Sedlak and Tom Kuhn deserve special thanks for encouraging me in this project even before I knew I would do it; Martin Krieger for cheering me on when it was too late to stop. My brother, Maury Fox, good-naturedly allowed me to raid his library, tolerated my many questions and even our disagreements.

I began the research for this book under a grant from the New York State University Research Foundation, wrote most of the first draft while on sabbatical from SUNY College at

Purchase, with the support of an Exxon Fellowship from the Program in Science, Technology, and Society at M.I.T. Don Blackmer and Carl Kaysen generously arranged for my continuing use of MIT facilities after the fellowship expired. Time for finishing the manuscript was made possible by a Mina Shaugnessy Award from the Fund for the Improvement of Postsecondary Education.

Archivists at the Rockefeller Foundation, Cornell University, California Institute of Technology, the Carnegie Institution of Washington, and the American Philosophical Society were among the many who helped me track down materials; others include Philip Alexander, Pnina Abir Am, Lee Ehrman, and Diane Paul. Linda Chaput and Betsy Dilernia at W. H. Freeman and Company saw to it that the manuscript became a book. And Lynn Robison suffered through the typing of its many drafts with patience and boundless good humor.

But finally, it was my children, Jeffrey and Sarah, who, by enduring my preoccupation and frustrations with unflagging love and support, helped the most to make this book possible.

A FEELING FOR THE ORGANISM

CHAPTER 1

A Historical Overview

The science of genetics as Barbara McClintock first encountered it was a very young subject—scarcely older than herself. Although Mendel's work had been rediscovered in 1900—two years before her birth—the term "genetics" had not even been coined until 1905, and "gene" was not a recognized word until 1909. Even then it was a word without a clear definition, and certainly without a material reality inside the organism on which it could be hung. At best it was an abstraction invoked to make sense of the rules by which inherited traits are transmitted from one generation to another. The suggestion that Mendelian "factors," or genes, are related to the chromosomal structures inside the cell, which cytologists studied, had been brilliantly argued by an American graduate student, Walter Sutton, in 1902, and independently by the German zoologist Theodor Boveri. But at that time it was an argument without direct empirical confirmation.

In the years before McClintock's arrival at Cornell in 1919, this connection began to acquire enough supporting evidence to make it compelling. Most of this evidence came from the work of a single laboratory, T. H. Morgan's "Fly Room" at Columbia University. Between 1910 and 1916, a series of chromosomal and genetic studies on the fruit fly *Drosophila*—by Morgan, A. H. Sturtevant, H. J. Muller, and C. B. Bridges—produced much of the evidence that was needed to confirm the relation between genes and chromosomes. The science of cytogenetics was born in this lab. Organisms with particular visible traits (in *Drosophila* these are mostly differences in eye color and wing shape) were crossed or mated, and researchers attempted to correlate the traits of successive generations with the inheritance of particular (X or Y) chromosomes. With these results, geneticists could confidently postulate a physical basis of Mendelian genetics. By 1915, the evidence was strong enough to permit Morgan, Sturtevant, Muller, and Bridges to publish their epoch-making book, *The Mechanism of Mendelian Heredity*, the first attempt to interpret all of genetics in terms of the chromosomal theory. The years that followed were years of heated controversy and avid promotional work on the part of T. H. Morgan. Even William Bateson, the early defender of Mendelian theory, resisted so "materialist" a basis for genetics, as indeed Morgan himself had earlier done. But as the evidence continued to accumulate, it became increasingly difficult to challenge the chromosomal interpretation. Still, most nonuniversity biologists, especially biologists in schools of agriculture, were somewhat less enthusiastic about the new biology than their university colleagues. To many of them, the Columbia research remained suspiciously "abstract."[1] Although Morgan had urged as early as 1911 that "cytology furnishes the mechanism that the experimental evidence demands," cytological work remained a relatively low priority in agricultural research.

In 1927, when Barbara McClintock received her Ph.D. in botany from the College of Agriculture at Cornell University, the infectious excitement generated by the marriage between cytology and genetics in Morgan's Fly Room at Columbia had not yet spread to Cornell. Probably the most important difference was that Cornell's geneticists studied corn, or maize, not fruit flies. Under the influence of Cornell's R. A. Emerson, the maize plant had become a powerful tool in the study of genetics. The colors of kernels on a cob of maize are a beautifully legible, almost diagrammatic expression of genetic traits. Whereas *Drosophila* offers the geneticist a new generation every ten days, maize matures slowly; the experimenter has time to gain more than a cursory acquaintance with each plant and to follow its development within a generation. But despite the extensive genetic studies, almost no chromosomal analysis of maize had yet been done. Barbara McClintock's work, begun while she was still a graduate student, demonstrated to her colleagues at Cornell that the genetics of maize, like that of *Drosophila*, could be studied not only by breeding the organism and watching the growing progeny but also by examining the chromosomes through the microscope. This new window into the mysteries of genetics was to prove crucial for the future development of genetics as a whole.

Using an important new staining technique that had just been developed by the cytologist John Belling, she succeeded in identifying and characterizing the individual chromosomes of maize by their lengths, shapes, and patterns. Once this was accomplished, she was in a position to integrate the results of breeding experiments (genetic crosses) with the study of chromosomes. In the years that followed, McClintock published a series of papers that elevated maize to a status competitive with *Drosophila*. At the same time, she established herself as one of America's leading cytogeneticists. In 1931, she and her student Harriet Creighton published a paper in the *Proceedings of the National Academy of Sciences* demonstrating that the exchange of genetic information that occurs during the production of sex cells is accompanied by an exchange of chromosomal material. It was called "A Correlation of Cytological and Genetical Crossing-over in *Zea mays.*" With this work, which had been referred to as "one of the truly great experiments of modern biology,"[2] the chromosomal basis of genetics was finally, and incontrovertibly, secured. Introducing it in his *Classic Papers in Genetics*, James A. Peters writes:

> Now we come to an analysis that puts the final link in the chain, for here we see correlations between cytological evidence and genetic results that are so strong and obvious that their validity cannot be denied. This paper has been called a landmark in experimental genetics. It is more than that—it is a cornerstone.[3]

Through the 1930s, at Cornell, at California Institute of Technology, and then at the University of Missouri, McClintock continued experiments and publications that consolidated, and complicated, the relation between cytology and genetics. She was elected Vice President of the Genetics Society of America in 1939, became a member of the National Academy of Sciences in 1944, and was President of the Genetics Society in 1945.

In the year of her election to the National Academy, she began the series of experiments that led her to transposition—work that many now see as the most important of her career. At the time, however, only she thought so. To most, her conclusions seemed too radical. But if 1944 was a crucial year in McClintock's career, it proved to be a pivotal year in the history of genetics as well—for reasons that had nothing to do with McClintock. It was the year in which the microbiologist Oswald T. Avery and his colleagues Colin MacLeod and Maclyn McCarty published their paper demonstrating that DNA provides the material basis for inheritance.

McClintock had begun her career in the midst of a major revolution in biological thought; now she was to witness a second, and equally momentous, revolution. The story of the birth of molecular biology has been told many times by now. It is a story of intense drama, fast action, colorful personalities, and high stakes. By the mid-1950s, molecular biology had swept the biological world by storm. It appeared to have solved the problem of life. It brought to biology a different world of inquiry, and a different model of scientific explanation. In this world of inquiry, McClintock's work came to seem increasingly idiosyncratic and obscure.

In 1938, Max Delbrück, a physicist who had turned his attention to the problem of heredity, had argued for the importance of bacteriophage—subcellular and submicroscopic particles only recently identified as a form of life—as "ideal objects for the study of biological self-replication."[4] In the summer of 1941, Delbrück arranged to meet Salvador Luria at the Cold Spring Harbor Laboratory, beginning a collaboration and a tradition that were to assume historic importance. Four years later, he organized the first summer phage course, "to spread the new gospel among physicists and chemists."[5] The heart of Delbrück's program was "the quest for the physical basis of the gene"[6]—not simply for its physical locus, the chromosome, but for the actual physical laws (and molecular structure) that constitute and explain the genetic mechanism. This quest was realized in 1953 with the discovery by James Watson and Francis Crick of the structure of DNA. From its structure, Watson and Crick were able to deduce how the DNA could perform the functions essential to genetic material—namely, replication and instruction. It was a euphoric time. According to Watson and Crick the vital information is coded into DNA, or, as it came to be known, the "mother molecule of life." From there it is copied onto the RNA, a kind of molecular go-between. Through a miraculously well-ordered physico-chemical process, the RNA is then used as a blueprint for the production of the proteins (or enzymes) responsible for genetic traits. The picture that thus emerged—DNA to RNA to protein—was powerful, satisfying, and definitive. Francis Crick dubbed it the "central dogma," and the name took. For the next ten years, biologists continued to experience the kind of growth and excitement that comes only in periods of scientific revolution.

In many ways, the picture that had thus been sketched bore strong resemblances to the picture of the universe that the Newtonian physicists had drawn. Both were highly mechanistic accounts in which, it seemed, only the details were missing. In each the essential principles were articulated in terms of the simplest possible system—for physicists this had been the interaction of two point masses; for biologists, it was the smallest and simplest living organism, the bacteriophage or, next best, the bacterium. For almost all workers in molecular biology, the bacterium to study was *Escherichia coli.* From *E. coli* to the rest of the living world was, presumably, a small step. Jacques Monod, the French Nobel Prize winner, reportedly said that what was true for *E. coli* would be true for the elephant.[7] A few biologists—Barbara McClintock was one—continued to study higher organisms, but the brightest and most promising young workers tended to take up the study of phage and bacteria. The language of corn genetics, once a staple in the training of all biologists, rapidly became unintelligible.

Given the confidence that many biologists had in the essential completeness of their understanding, it seems inevitable that at some point they would come to feel that the interesting questions had all been answered—that what remained was hackwork. In the late 1960s, many prominent molecular biologists were looking for new fields in which to reinvest their energies and talents.

As it turned out, the anxiety (or satisfaction) that these biologists had felt over the exhaustion of their subject was premature. What was true for *E. coli* was *not* true for the elephant; as would emerge later, it wasn't even always true for *E. coli.* As has so often occurred in the history of science, just when faith in the nearness of the goal was at its greatest, increasingly vexing and disturbing observations began to accrue. Over the next ten years, developments would emerge that would vastly complicate the picture that had been so simple—that would, in the minds of many, radically challenge the central dogma.

The principal thesis of the central dogma as originally articulated by Crick was that "once 'information' has passed into protein *it cannot get out again.*"[8] Information flowed *only* from DNA to RNA to protein. Crucial to this thesis was that information originated in the DNA and that it was not then subject to modification.

In its original form, the central dogma offered no way to account for the fact that the specific kinds of proteins produced by the cell seemed to vary with the cell's chemical environment. By 1960, a crucial emendation was made by Jacques Monod and François Jacob. Monod and Jacob worked out a mechanism for genetic regulation that allowed for environmental control of the *rates* of protein production but retained the basic tenet of the central dogma. Proteins, they assumed, are encoded by the DNA, but so are a number of switches, each of which can turn "on" and "off" a gene (or group of genes, called structural genes) encoding the proteins. According to their theory, the switch itself is made up of two different kinds of genetic elements, an "operator" and a "regulator" gene, which, in concert with a particular chemical substrate, repress or activate the structural gene. The availability of the relevant chemical substrate is in turn determined by the chemical milieu of the cell.

Thus modified, the central dogma was now stronger than ever, with a domain that was vastly enlarged. It seemed to have resolved and incorporated, at least in principle, the troublesome field of enzymatic control. True, information could now "get out of the protein," but only in the form of the regulation of the rate at which a particular stretch of the DNA would be processed. The feedback that had been introduced left the most essential feature of the central dogma intact: the information flow from the DNA remained one way. For this work, Jacob and Monod shared the Nobel Prize with Andre Lwoff in 1965.

On the other side of the Atlantic, Barbara McClintock had been trying since 1950 to interest biologists in her own identification of "controlling elements" in maize. When Monod's and Jacob's paper appeared in *Comptes Rendus*, late in 1960, she was one of its most enthusiastic readers. With great excitement and in full support of her French colleagues, she promptly called a meeting at Cold Spring Harbor to outline the parallels with her own work. Shortly after, she detailed these parallels in a paper, which she sent first to Monod and Jacob and then to the *American Naturalist.* As Horace Judson reports in *The*

Eighth Day of Creation, his account of the molecular revolution in biology, "they were glad of her prompt support."[9]

McClintock's investigations were conducted not only on a different continent, not only on a vastly more complex organism, but, it could be said, in a different biological world from that of Monod and Jacob. Where Monod and Jacob, as molecular biologists, concentrated their efforts on *E. coli,* McClintock, a classical geneticist, studied maize. Where Monod and Jacob employed the tools of biochemical assay to determine the effects of critical genetic crosses, McClintock used the techniques more familiar to the naturalist—she observed the markings and patterns of colorations on the leaves and kernels of the corn plant, and the configurations of the chromosomes as they appeared under the microscope. Where they sought a molecular mechanism, she sought a conceptual structure, supported and made real by the coherence of its inferences and its correlation with function. Each used the tools, techniques, and language appropriate to his or her organism, milieu, and time. Begun in 1944, McClintock's work belonged in the tradition of classical biology; it was, above all, premolecular. Its beginning preceded even the identification of DNA as the genetic material. By contrast, Monod and Jacob worked in a spirit that belonged entirely to the age of molecular biology, the age of the central dogma. Yet they, like McClintock, had become convinced of the existence of regulatory mechanisms operating at the genetic level, and involving two kinds of regulatory genes. The controlling element McClintock had identified adjacent to the structural gene (the gene directly responsible for the trait) appeared analogous to Jacob and Monod's "operator" gene; her "activator" gene, analogous to their "regulator" gene, might be independently located. As McClintock wrote in her paper on the parallels between maize and bacteria, in both cases, "an 'operator' gene will respond only to the particular 'regulator' element of its own system."[10]

But one essential feature was different. In McClintock's system, the controlling elements did not correspond to stable loci on the chromosome—they moved. In fact, this capacity to change position, *transposition* as she called it, was itself a property that could be controlled by regulator, or activator, genes. This feature made her phenomenon a more complex one, and, in the minds of her contemporaries, less acceptable. Although it was known that viral DNA could insert itself into the DNA of a host cell and subsequently detach itself, almost no one was ready to believe that, under certain circumstances, the normal DNA of a cell could rearrange itself. Such a notion was upsetting for many reasons, not the least of which was that it implicitly challenged the central dogma, which during the 1950s and 1960s grew ever more entrenched. If parts of the DNA might rearrange themselves in response to signals from other parts of the DNA, as McClintock's work suggested, and if these signals might themselves be subject to influence by the internal environment of the cell, as the regulatory elements of Jacob and Monod clearly were, what then becomes of the unidirectionality of the information flow from DNA to protein? For the very sequence of genes to depend on factors beyond the genome, information would, in some sense, have to flow backward, from protein to DNA. McClintock did not make this suggestion explicit, but, with her interpretation, the organization of the maize genome clearly became a much more complex one than the central dogma would allow.

To many biologists in the 1950s and 1960s transposition sounded like a wild idea. Furthermore, fewer and fewer knew enough maize genetics to follow the very intricate arguments that were necessary to support the radical conclusions McClintock had reached. The mood in biology had grown impatient with the complexity of higher organisms. Finally, it must be said, her own writing was dense. The reprinting of her early paper with Harriet Creighton in Peters's *Classic Papers in Genetics* is introduced with the warning:

> It is not an easy paper to follow, for the items that require retention throughout the analysis are many, and it is fatal to one's

> understanding to lose track of any of them. Mastery of this paper, however, can give one the strong feeling of being able to master almost anything else he might have to wrestle within biology.[11]

The newer papers were even more difficult to follow. When she made these ideas public in 1951, in 1953, and again in 1956, in spite of the fact that she had long since established her reputation as an impeccable investigator, few listened, and fewer understood. She was described as "obscure," even "mad." Even in 1960, when she outlined the close parallels between the system of Jacob and Monod and her own, very little attention was paid. Jacob and Monod themselves neglected to cite her work in their major paper on regulatory mechanisms of 1961—"an unhappy oversight," they later called it.[12] They did, however, acknowledge her work in a concluding summary they wrote for the Cold Spring Harbor Symposium the next summer:

> Long before regulator and operator genes were revealed in bacteria, the extensive and penetrating work of McClintock . . . had revealed the existence, in maize, of two classes of genetic "controlling elements" whose specific mutual relationships are closely comparable with those of the regulator and operator. . . .[13]

No mention was made there of transposition, although elsewhere in the same volume they remarked that the occurrence of transposition in maize made for "an important difference between the two systems."[14]

Ten years later, when in many people's expectations molecular biology should have run its course, a number of dramatic and unanticipated observations began to appear. Among these was the startling discovery of elements of the bacterial genome that appeared to "jump around." These were variously called "jumping genes," "transposons," or "insertion elements." In a number of cases, transposons have been observed to have regulatory properties that closely parallel those observed twenty years earlier by McClintock; indeed, they seem to present a closer analogue for her controlling elements than that of Monod and Jacob's regulatory system. The major difference is that they are described in a new language, the language of DNA.

In the decade since the first discovery of jumping genes in bacteria, it has become apparent that the movement of genetic elements can play a critical role in some quite ingenious regulatory mechanisms—both more complex and more varied than those that had been envisioned by Jacob and Monod. For example, in some instances, the regulatory function may depend on the orientation—forward or backward—with which the transposed element is reinserted; the two orientations correspond to functionally different genetic scripts. Apart from maize, yeast was the first eukaryotic system to attract attention to the distinct developmental changes that result from a controlled rearrangement of genes; now a number of developmental mutations in *Drosophila* have also been traced to transposition. Even in mammals a kind of genetic mobility can be seen. Studies of the immune system in mice have revealed that immunoglobulin DNA is routinely subject to extensive rearrangements, offering a possible explanation of the heretofore inexplicable diversity of antibodies.

Cold Spring Harbor today is a center for much of the new research on transposition. And Barbara McClintock, who has lived and worked at Cold Spring Harbor since 1941, continues to do her own work—still in relative seclusion. For forty years, she has been sheltered from the vicissitudes of biological fashion by an aura of privacy and reserve. Until recently, her following has been small, reverent, and protective.

For those few who have sought to know the person and work behind the name, and who have tried to understand how that work fits into modern biology, the knowledge they have come by is felt as a special privilege. One such enthusiast is Nina Federoff, of the Carnegie Institution of Washington, who has set her sights on an understanding of the molecular basis of McClintock's genetic analyses. For her, reading McClintock's papers became "the most remarkable learning experience" of her life. "Like a detective novel, I couldn't put it down."[15]

Barbara McClintock with Walter Gilbert, Harvard Commencement, Cambridge, Mass., June 1979. (Permission of Rick Stafford, photographer.)

Now, active public recognition and fame threaten to intrude on years of obscurity and reserve. Over the last few years, the number of honorary citations has begun to mount. In 1978, recognizing that "Dr. McClintock has never received the formal recognition and honor due such a remarkable scientist," Brandeis University chose her for their annual Rosenstiel Award, citing her "imaginative and important contributions" to the world of science. In 1979, she was awarded two honorary degrees—one at Rockefeller University and one at Harvard. The Harvard citation reads: "A scientific pioneer, firm-purposed and undaunted; her profound and pervasive studies of the cell have opened avenues to deeper understanding of genetic phenomena." The Genetics Society of America saluted her in 1980 "for her brilliance, originality, ingenuity, and complete dedication to research." In 1981, her name caught the eye of the public with the award of the first MacArthur Laureate Award—a lifetime fellowship of $60,000 a year, tax free. Her story became media news. A day later, she received the prestigious Lasker Award for Basic Medical Research and a $50,000 prize from Israel's Wolf Foundation (the ninth honor that year). The Lasker citation noted the "monumental implications" of her discoveries, which it said were "not fully appreciated until years later." When, in the fall of 1982, she shared Columbia University's Horwitz Prize with Susumu Tonegawa, commentators made note of the frequency with which past Horwitz Prize winners were subsequently named Nobel laureates.

For McClintock, all this attention is "too much at once." *Newsweek* described her, seated before a roomful of reporters, as "plainly miserable." "I don't like publicity at all. All I want to do is retire to a quiet place in the laboratory."[16]

But the question of how she regards the latest chapter in the history of her subject goes beyond personal satisfaction or discomfort. To what extent does she see in it the unfolding of a vision she has cultivated over more than three decades? How does her own account add to our understanding of the history of this period, indeed of the entire history of genetics? Understanding that the spirit of a scientific era cannot be learned from the scientific or historical literature alone, we need to know about the lives and personalities of the men and women who have made the science. In the chapters that follow, the history that has been told so briefly here is retold through McClintock's —and others'—own recollections.

CHAPTER 2

The Capacity to Be Alone

Forty miles east of Manhattan, along route 25A, just a mile before the town of Cold Spring Harbor, the sign for the Long Island Biological Laboratories is easy to miss. Small and inconspicuous, nothing about the turnoff to the laboratories suggests what a major thoroughfare Bungtown Road can be. At the beginning of each summer, when the pace of university laboratories slows down, biologists from all over the world gather here to work or study, or simply to meet and share the latest results of one another's work. Annual symposia draw an especially large crowd, straining the modest facilities beyond their capacity. Everywhere small groups congregate in animated talk, spilling onto the beach and roads. From June to early September, Bungtown Road seems to these biologists more like a central artery than the small country road it is.

Fewer people know Cold Spring Harbor in wintertime. When the excitement of summer conferences subsides, and the visiting scientists leave, only a few biologists remain. The weather turns cool, the small beach empties, the leaves begin to turn color. Then the resident scientists can return to their research, undisturbed by the intrusions that plague their colleagues at universities and urban research centers. Here there are no city lights to distract, nor university demands to disturb. Cold Spring Harbor offers, to those who make it their home, an atmosphere of peace and quiet few major laboratories know.

In the fall of 1978, I drove out to Cold Spring Harbor to get, on tape and on paper, a record of Barbara McClintock's musings and recollections. Many years before, I had spent a summer at the Long Island Biological Laboratories as a graduate student. I remembered seeing her—contained, aloof, perhaps even eccentric—going to and from her laboratory or on solitary walks in the woods or by the beach. But I never talked to her or visited her lab even though I briefly worked in the very same building. Her domain was set apart from the molecular biologists I was studying with by a wide gulf and, at the time, like most others, I was not even curious. When I returned twenty years later, the setting of the laboratories seemed idyllically quiet—the cool breeze of autumn just beginning, the sun and blue waters still brilliant. It summoned up a different image from the summertime Cold Spring Harbor I had known as a student. It was the image of retreat, of romantic seclusion—where one's work might be everything.

I found McClintock in her lab, although the word is inadequate to describe what felt more like a world. Certainly it was the most lived-in laboratory I had ever been in. Adjacent to her lab and tucked away in the far end of a large concrete building that was named after her in 1973, her office is notable only for its simplicity. An inlet from Long Island Sound comes to within a stone's throw of her window. She herself seemed barely to have changed in the years since I'd seen her last, in spite of a little more grey in her short-cropped hair, a few more lines in her face. Her slacks and shirt pointedly rejected feminine fashion, but they were carefully pressed. The economy of her words and movements, the way she dressed, the way she moved and talked—all expressed a fastidious spareness, an aesthetic of order and functionality, that seemed to defy age.

During the ride out, I'd recalled with some apprehension how this person had been described to me: "intimidating," "difficult to approach," "a great mind," "penetrating and demanding," "a very private person." In fact, she greeted me with surprising warmth, engaging me immediately in direct personal exchange. At the same time, there was no doubt about who was in charge: she seated me in the easy chair behind her desk, and herself in the straight-backed chair facing me; I would be the first to be interviewed. I gave the best account I could of myself, my background, and my interests. And we talked. Within minutes, we were deep in a first conversation that was to last five hours. We talked about women, about science, about her life—but she did not want to be interviewed.

She did not see how her life could be of possible interest to the world. Certainly she did not think that her experience could be of any particular value to women. On this she was adamant; she was too different, too anomalous, too much of a "maverick" to be of any conceivable use to other women. She had never married, she had not, as an adult or as a child, ever pursued any of the goals that were conventional for women. She had never had any interest in what she called "decorating the torso." Gradually, with effort, I was able to persuade her to let me record her version of her life. My argument was that her story was important precisely because it was so unconventional.

Barbara McClintock has lived most of her life alone—physically, emotionally, and intellectually. But no one who has met her could doubt that it has been a full and satisfying life, a life well lived. Perhaps the word that best describes her stance is "autonomy." Autonomy, with its attendant indifference to conventional expectations, is her trademark. Where had this extraordinary "capacity to be alone" (in the phrase of the psychoanalyst D. W. Winnicott[1]) come from?

• • •

To some extent it might be said that she was born into the role of maverick and pioneer. With a mother descended from solid Yankee stock and a father born to Celtic immigrant parents, rugged individualism might well be part of her self-expectations.

Barbara's mother, Sara Handy McClintock, was an adventurous and high-spirited woman. She was the only child of Sara Watson Rider and Benjamin Franklin Handy—a marriage joining two of Hyannis's most venerable families, both of them tracing their lineage back to the Mayflower. Sara (christened "Grace") was born on January 22, 1875. Her mother died before the year was out, and Sara was taken to California to be cared for by an aunt and uncle who had been lured out there by the Gold Rush of 1849. But in both fact and imagination, Hyannis remained her parental home. Years later, when Sara Handy was a mature woman with four children of her own—a member of The Society of Mayflower Descendents and a Regent of the Battle Pass Chapter of the Daughters of the American Revolution—she romanticized the family history in a small privately published volume of light verse.

With considerable nostalgic charm, these poems convey a vivid picture of the New England life of her seafaring ancestors. The most colorful of these is Hatsel Handy, Sara's grandfather, who shipped out to sea at the age of twelve, and by nineteen was already captain of his vessel. "Cap'n" Handy is portrayed as a fun-loving, independent-minded adventurer with a wry wit. The book's epigraph, "Don't it beat all how people act if you don't think their way," was Sara's own motto, lived out by her grandfather if not by her father.

Benjamin Handy was a stern and righteous man, a Congregationalist minister, and even at a distance he exerted a strong influence on the rearing of his daughter. Back from California, Sara Handy grew into an attractive and highly eligible young woman: an accomplished pianist, an amateur poet and painter as well. But when it came time to think about marriage, she had to contend with the harsh judgments of her father. Benjamin Handy tended to disapprove of all his daugher's suitors, and vehemently so in the case of Thomas Henry McClintock. Thomas Henry was born in Natick, Massachusetts, in 1876; the son of immigrant parents from the British Isles: a foreigner by

Handy standards. Furthermore, he was still in medical school, and hardly in a position to support a family. But he was an attractive, determined young man who made a great impression on the high-spirited and independent Sara. Against her father's wishes, they married in 1898, shortly before young McClintock's graduation from Boston University Medical School.[2]

From her own small inheritance, Sara helped her husband pay his medical school debts. Then, without any support from Reverend Handy, they established a home—first in Maine, then New Hampshire, and soon after in Hartford, Connecticut—and, in rapid succession, a family. Marjorie was born in October 1898, Mignon in 1900, and Barbara on June 16, 1902. A son, christened Malcolm Rider but always known as Tom, was born a year and a half later.

• • •

Sara McClintock holding son, Malcolm Rider ("Tom"), with daughter, Barbara, and Uncle Lloyd Hanson Rider, in Hartford, Conn. (Permission of Marjorie M. Bhavnani.)

By McClintock's own account, her "capacity to be alone" began in the cradle: "My mother used to put a pillow on the floor and give me one toy and just leave me there. She said I didn't cry, didn't call for anything." Her temperament, she says, led her parents to change her name when she was only four months old. Instead of Eleanor, a name they had originally chosen as especially feminine and delicate, they soon decided that "Barbara" would be more appropriate for a girl with such unusual fortitude. It sounded to them more masculine. The family tradition says that they had hoped for a boy, to be named Benjamin like his grandfather.

By Marjorie McClintock's account, their mother was under acute stress when her third child was born. She was ill-prepared by her own relatively privileged rearing for the hardships of raising four small children with little or no help. It took a number of years before Dr. McClintock was able to establish himself in medical practice, and, in the meantime, money was scarce. His wife contributed by giving piano lessons, trying at the same time to protect a space for her own artistic interests. Dr. McClintock in turn helped as he could ("he loved babies"), but the arrival of the fourth child taxed the mother's strength to the limit, and Barbara bore the brunt of the strain. Relations between Barbara and her mother were marked by tension from the very beginning, and little more than a year after Tom's birth, Barbara was sent to live with her paternal aunt and uncle, in Massachusetts. She recalls the periods she spent with them, on and off until she reached school age, with warm enthusiasm: "I enjoyed myself immensely." With pride, she asserts she was "absolutely not" homesick.

Her uncle was a fish dealer, and she especially enjoyed trips with him and his horse and carriage first to the fish markets, and then out in the country. "A great big man with a great big voice, [he] used to yell out, 'Do you want any fish?' Then the housewife would answer and come out." Later her uncle bought a motor truck, and McClintock dates her early interest in motors back to watching him struggle with the constant breakdowns of his new machine.

Barbara McClintock as a child, circa 1907. (Permission of Marjorie M. Bhavnani.)

Machines, tools, and the skills of mechanics were a bond with both uncle and father. "My father tells me that at the age of five I asked for a set of tools. He did not get me the tools that you get for an adult; he got tools that would fit in my hands, and I didn't think they were adequate. Though I didn't want to tell him that, they were not the tools I wanted. I wanted *real* tools, not tools for children."

When Barbara was back at home her relationship with her mother became more distant than ever. A resounding "No!" greeted her mother's attempts to embrace her. The tensions between them seemed only to increase, and she speculates that they may have been responsible for her unusual self-sufficiency. In any case, according to the family legend, Barbara grew up as a solitary and independent child.

In 1908, the family moved to a part of Brooklyn, New York, that was still semirural. The children attended the local elementary school and, later, Erasmus Hall High School. By this time things were a little easier, and the family was able to spend summers at the far end of Long Beach—then still wild and unspoiled country. Both parents, Barbara recalls, wanted the children to learn to feel free in the water. "I remember getting up early in the morning and walking with the dog. I used to love to be alone, just walking along the beach." Evenings, when no one was around, she would go out and run in a special style that she had discovered herself: "You stood quite straight with your back just completely straight, and you practically floated. Each step was rhythmically floating, without any sense of fatigue, and with a great sense of euphoria." Years later, reading Margaret Mead, she learned that her secret had been known by others; there were Buddhist monks in Tibet, "Running Lamas," who cultivated the same technique.

The experiences she valued most as a child were solitary ones. She was an avid reader, and, best of all, she loved to sit alone, intensely absorbed, just "thinking about things." All of this sitting alone worried her mother: "She felt there was something wrong," Barbara remembers. "I knew there was really nothing wrong; my sitting there was related to things that I was thinking about." But at the same time, her mother was also manifestly appreciative of Barbara's musings. She kept a notebook, recording some of Barbara's ideas. Barbara recalls only one of the early scientific explanations her mother noted down: "My mother was crushing strawberries for strawberry shortcake (I don't know how old I was at this time, but I was quite young), and I was watching her. 'Now I know where blood comes from,' I said, 'It comes from strawberries!' "

Barbara loved music, but piano lessons with her mother were soon discontinued, for she applied herself to the instrument with a painful intensity that Mrs. McClintock felt could not be good for her. With another teacher things were no better, and piano lessons stopped altogether. "This intensity or this sense of feeling disturbed about situations, or taking them too difficultly, led me to be taken out of school on several occasions."

In general, all the children were encouraged in their own interests, and most of the time, for both parents, the children's preferences were overriding. If Barbara didn't want to go to school, then she didn't go; the same was true for her siblings. Sometimes she was taken out for long periods—as long as a semester or more. She recalls one time in particular when she had come to dread going to school because the teacher disturbed her so: "She seemed to be ugly, not only physically ugly, but ugly in her character, ugly in her emotions." Though she realizes now that her picture was a caricature, "I can still see that caricature, I can still see it in action." Given the intensity of her feelings, her parents decided that she shouldn't even try to go to school. ("My father should have been a pediatrician because he was very perceptive with children.")

In Barbara's family, school was regarded as "just a small part of growing up." From the beginning, Barbara's father took an exceptional stand: he made it clear to the school officials that his children were not to be given homework—six hours a day of school were more than enough. According to Marjorie, the parents' concern was not with what the children *should* be, but with what they were. When her parents noticed that Barbara liked to ice skate, she remembers, "they bought me the very best skates, the very best shoe and skate outfit that they could get for me. Every good day for skating, I was out skating at Prospect Park rather than going to school." When school let out and her brother and his friends came home, there was the world of street sports—of baseball and football and volleyball and teams. For these things, she needed the appropriate clothes.

"At that time," she recalls, "one didn't go to a store to buy one's clothes. We had a dressmaker come in for the girls and had our clothes made." Very early she insisted, and persuaded her parents ("They always did acquiesce to anything that I wanted that they thought was important to me") that she should have bloomers made of the same material as her dress —"so that I could do anything that I wanted. I could play baseball, I could play football, I could climb trees, I could just have a completely free time, the same kind of a free time that my brother and the people on the block had." She remembers having no girlfriends at this time, only boyfriends.

"One time when I was out playing basketball or volleyball or something, a woman on the block called me to her house, and I went up the steps leading to the front door. She invited me in, stating that it was time that I learned to do the things that girls should be doing. I stood there and looked at her. I didn't say anything, but I turned around and went directly home, and told my mother what had happened. My mother went directly to the telephone and told that woman, 'Don't ever do that again!' "

It wasn't that Barbara was felt to be *like* the rest of the family —on the contrary, both her mother and her father felt her to be quite different. On the other hand, she was not felt to be *more* individualistic than the others: "Barbs was simply Barbs," as her sister puts it. Her parents were willing and even anxious to protect these differences—at any rate up to adolescence. Then, when she didn't outgrow her tomboyish ways and "become like other girls," and especially when she began to show "intellectual desires," her mother at least became anxious about her future.

Major Thomas Henry McClintock, France, 1918. (Permission of Marjorie M. Bhavnani.)

Throughout adolescence it became increasingly clear that she was committed to "the kinds of things that girls were not supposed to do." The passion for sports gave way to the passion for knowledge. "I loved information," she remembers. "I loved to know things." At Erasmus Hall High School, she discovered science. It was there that the pleasure of solving difficult problems began to grow. "I would solve some of the problems in ways that weren't the answers the instructor expected. . . . I would ask the instructor, 'Please, let me . . . see if I can't find the standard answer' and I'd find it. It was a tremendous joy, the whole process of finding that answer, just pure joy."

• • •

When World War I began and the National Guard was called up, Dr. McClintock was sent overseas as a military surgeon. Hard times returned, and Mrs. McClintock was obliged to take on more piano teaching. During most of the children's adolescence their father was away, and a number of crucial decisions had to be made by their mother alone. There were choices to be made for each of her children, but those for her youngest daughter were perhaps the most difficult of all.

Marjorie and Mignon were both unusually successful students in high school, and Marjorie was offered a scholarship at Vassar. But being concerned about money and fearful that too much education was likely to make a young woman unmarriageable, Mrs. McClintock was able to dissuade her oldest daughter from going to college. Marjorie became a professional harpist and, like her mother, a gifted pianist. Both girls married, after flirting briefly with the idea of a career on the stage. Marjorie played with the Washington Square Players; Mignon went on tour with a group to Chicago.

When it came to her two younger children, however, Mrs. McClintock's influence failed. Tom, the only son, left the family sphere altogether; in the tradition of his grandfather, he ran off to sea while still in his teens. Barbara's interests took her in even less acceptable directions. As her commitment to learning grew stronger, so did Mrs. McClintock's concerns about her daughter's future. Barbara recalls, "She was even afraid I might become a college professor." She worried that Barbara would

Sara McClintock, Malcolm Rider ("Tom"), Barbara, Mignon, and Marjorie, circa 1918. (Permission of Marjorie M. Bhavnani.)

become "a strange person, a person that didn't belong to society"; she must also have known that becoming a college professor would not be easy.

Barbara, too, was beginning to realize that her course would be difficult. Later she would often recall an incident from this early period with a sense of particular significance: "We had a team on our block that would play other blocks. And I remember one time when we were to play another block, so I went along, of course expecting to play. When we got there, the boys decided that, being a girl, I wasn't to play. It just happened that the other team was minus a player, and they asked me would I substitute. Well we beat our team thoroughly, so all the way home they were calling me a traitor. Well, of course, it was their fault. But I understood at the time that I just had to make these adjustments to the fact that I was a girl doing the kinds of things that girls were not supposed to do."

During high school, Barbara found she had to think through "how I could handle the fact of my difference." Reluctantly, she acknowledges that this process was not easy. "I found that handling it in a way that other people would not appreciate, because it was not the standard conduct, might cause me great pain, but I would take the consequences," she says. "I would take the consequences for the sake of an activity that I knew would give me great pleasure. And I would do that regardless of the pain—not flaunting it, but as a decision that it was the only way that I could keep my sanity, to follow that kind of regime. And I followed it straight through high school, and through college, through the graduate period and subsequently. It was constant. Whatever the consequences, *I had to go in that direction.*"

• • •

Although Barbara's parents did not directly support her interest in science, or, for that matter, in the life of the mind, they did provide a precedent that was perhaps even more important: their profound respect for self-determination. With the exception of her mother's strongly voiced misgivings about college education for women, pressure to conform to social expectations was almost nonexistent.

As her mother had done before her, then, she overrode parental opposition, and in 1919 she made her way to Cornell University to enroll in the College of Agriculture. Although it seemed a novel move to her family, in a larger context it was by no means exceptional. Women had been moving in from the periphery of American letters and science since the early part of the nineteenth century. By the turn of the century, the movement for higher education for women of her class and background was in full swing. Five women's colleges flourished in New England alone, and a number of major universities had become coeducational.

The women who took advantage of the new opportunities open to them were, for the most part, upper-class and upper-middle-class women. A disproportionate number of them came from English and Celtic families that were usually from New England. Many of these women chose to study science.

More than half a century earlier, Maria Mitchell had been elected to the American Academy of Arts and Sciences after her discovery of a new comet. Others after her (among them, Lydia Shattuck, Annie Jump Cannon, Cornelia Clapp, Ellen Swallow Richards, and Nettie Stevens) established further precedents. Under Mitchell's leadership, a movement arose in the 1870s to encourage women to pursue the study of science and to help them enter scientific occupations. By 1920, the porportion of women trained in science was at an all-time high. Fifty years later, when their representation had dropped to less than half what it had been earlier, women could look back at this period with surprise and a little envy.

Apart from the women's colleges, two universities stand out as having been especially hospitable to women students in the sciences during this period. They were the University of Chicago and Cornell. Cornell had been founded on the commitment to an education for "any person in any study." The first woman was admitted in 1872, and in 1873, the building of Sage College was begun. At the laying of the cornerstone, its benefactor, Henry W. Sage, had prophesied: "The efficient force of the human race will be multiplied in proportion as woman, by culture and education, is fitted for new and broader spheres of action."[3] In return for this gift, Cornell pledged itself to "provide and forever maintain facilities for the education of women as broadly as for men."[4]

Even though practice and vision did not fully coincide, by the early part of the century, Cornell had begun to attract a large number of highly motivated young women who pursued the intellectual life with vigor and success. In Barbara McClintock's graduating year, 1923, 74 of the 203 bachelor of science degrees that Cornell granted went to women. These figures include degrees from the Colleges of Home Economics and Hotel Management, but they are nonetheless impressive. Approximately 25 percent of the graduates from the College of Agriculture alone were female. In the Arts College, where the ratio of men to women was roughly four to one, two-thirds of those who graduated with honors were women. In that same class, more than half of the undergraduate scholarships had been awarded to women, as well as a substantial number of the graduate fellowships. A large proportion of these awards were in the sciences: in physics, mathematics, and biology.

McClintock does not remember where she first heard about Cornell or how she came to set her heart on going there to college. But decide she did, quite early, that she would go to college and that she would go to Cornell. (Her sister reminisced, with affectionate pride, that when Barbara *had* to do something, that was all there was to it.) But her mother's misgivings had by this time grown into active opposition, and her father was still overseas in the army. The shortage of money was acute. Barbara seemed to be defeated. Graduating from high school a semester early, in 1918, she went to work at an employment agency. For six months (she was still sixteen), she spent her days interviewing people to place them in suitable jobs; after work her afternoons and evenings were spent in the library getting an education. "I had a schedule—things I'd read—and I was going to educate myself some way or other," she says. "I was going to get the equivalent of a college education if I had to do it on my own."

Toward the end of the summer, her father came back from Europe and apparently threw his weight onto Barbara's side of the argument. "One morning . . . before I went to work, my mother said that she and Dad had been talking about my situation and had decided I was to go to college. So she called up a friend who went to Cornell and asked when Cornell opened. Well, this girl said that Cornell opened the following Monday, and that the 'Ms' were to register at 8:00 Tuesday morning. I went to work, and Mother went to my high school to see about credentials and so forth, but she brought nothing back with her. Nevertheless, I went up on Monday on the train to Ithaca, got myself a place to stay in a rooming house, and at 8:00 on Tuesday I went in a line with everybody else whose name began with 'M.' Everybody else had papers, but I didn't have anything but me, and the registrar said to me when I got there, 'You haven't got anything. How do you expect to get in?' And just at that moment my name was called out in the room, loud enough for both of us to hear. He said, 'Wait a minute,' he walked back, talked to somebody, came back, said 'Take the papers, go ahead.' I never found out what happened, haven't the foggiest notion what happened. But from then on I was in. Somebody did something, and I don't know what. It was just one of the coincidences. All I cared about was that I was in. And I was entranced at the very first lecture I went to. It was zoology, and I was just completely entranced. I was doing now what I really wanted to do, and I never lost that joy all the way through college."

The sense of things "just happening" was evidently something that Barbara enjoyed; it gave support to her notion that different rules applied to her. Her sister, however, is certain that beyond the apparent mystery lay the ingenuity of her mother. "Mother was most resourceful, and would have done everything necessary to see that Barbara was admitted, once the decision was made."

Cornell was everything Barbara had hoped for. At the College of Agriculture, tuition was free (a crucial factor in the final decision to enroll), but, even so, money continued to be a source of worry. Partly out of uncertainty about how long her money would last, and partly out of zeal, she signed up intially for a large overload of courses. "I would sign up for a course, but if I thought it was dreadful, I'd drop out, which would give me a 'Z.' By the time I was a junior, I discovered I had quite a lot of 'Zs.' "

This behavior was to cause the registrar's office some headaches, and it ultimately prevented what would otherwise have been an early graduation, but it was just that kind of institutional formality with which, then as later, Barbara could not or would not be bothered. Life at Cornell was simply too interesting.

"There were many things that one learned in college that one could not ordinarily learn then outside. You met people from all kinds of groups and societies; you were able to gain knowledge from people from different places, with different backgrounds. . . . College was just a dream. . . . I got to know many people at Cornell, and I got in with girls who were mainly

Jewish—at that time there was a great separation between the Gentiles and Jews. I liked this group very much. We were living in dormitories, and two of them were roommates. They had two rooms living in a tower and that used to be our meeting room." She was drawn to them—even to the extent of spending time learning to read Yiddish—"because they were quite different from the rest of the population at Cornell."

Barbara McClintock at her parents' home on Rutland Road, Brooklyn, New York, in the early 1920s. (Permission of Marjorie M. Bhavnani.)

This group (Emma Weinstein, later a leading figure in the Jewish community of New York, was a member; Laura Hobson, author of *Gentleman's Agreement* and one of the few non-Jewish members, another) evidently constituted a social core for Barbara. She was the only scientist in the group.

But if Barbara was drawn to this "separate" group, it was not because she herself felt isolated from her classmates in general. In her freshman year, her sister Marjorie recalled, she "blossomed" socially, in dramatic contrast to her solitary childhood. She was "lovely to look at," she received invitations—when she came home her mother and sisters were "just delighted." So successful was her involvement with the life of the school that in her first year she was elected president of the women's freshman class. She was also rushed by a sorority, but when she discovered that she was the only one in her rooming house to be invited to join, she declined the invitation. "Many of these girls were very nice girls, but I was immediately aware that there were those who made it and those who didn't," she remembers. "Here was a dividing line that put you in one category or the other. And I couldn't take it. So I thought about it for a while, and broke my pledge, remaining independent the rest of the time. I just couldn't stand that kind of discrimination. It was so shocking that I never really got over it; even now I feel very strongly against honorary societies. I belong to a number, because I have to if I have a job. If I hadn't had a job, I could afford to say no. But I have to abide by the rules of a job. I just don't go to meetings. . . . But I have to join." Her sister Marjorie feels that it may have been this reaction that broke the spell of Barbara's first year and tipped her back into declared nonconformity.

Many aspects of the life she could or could not, would or would not live were being worked out in those years—all with the same singlemindedness. She decided, for example, as an undergraduate, that she could no longer be bothered with long hair. She and the local barber had a "long philosophical talk" about it, and off came her hair. Her new haircut "made quite a noise on campus the next day—all over the campus." Shingled hair became fashionable soon after that, but in 1920 "it just happened to be a little ahead of something that was coming anyway!" By the same personal reasoning, a couple of years later when she was a graduate student, she decided that she could not work in the cornfields wearing the dresses and skirts that other women wore. Off she went to the tailor and ordered a pair of knickers, or "plus fours," as they were then called. It was a question of "what I could live with, and I couldn't live with the costumes that others had in the past."

But perhaps the single most important issue to be resolved then was her relation to men. In her first two years, she had lived the life of a coed, going out on lots of dates. "Then I finally decided I had to be more discriminating. I remember being emotionally fond of several men, but they were artists of one form or another, they were not scientists. They were not casual involvements. But when I felt a very strong emotional attachment, it was an emotional attachment, nothing else." On this she was very clear, as she was about its implications: "These attachments wouldn't have lasted. I knew [with] any man I met, nothing could have lasted. I was just not adjusted, never had been, to being closely associated with anybody, even members of my family. . . . There was not that strong necessity for a personal attachment to anybody. I just didn't feel it. And I could never understand marriage. I really do not even now. . . . I never went through the experience of requiring it."

On the other hand, it was not the vision of a chosen career that drew her instead. Nothing in McClintock's life corresponded to the kind of deliberate planning such a choice would imply. "At no time had I ever felt that I was required to continue something, or that I was dedicated to some particular endeavor," she says. "I remember I was doing what I wanted to do, and there was absolutely no thought of a career. I was just having a marvelous time."

But whether by calling or by circumstance, by the end of her junior year, she was well on her way to becoming a professional scientist. When I asked about the kinds of antecedents that are familiar in the lives of so many male scientists, she didn't seem to be aware of many. No one else in her family was interested in science, and though her father was a doctor, and had been especially devoted to her (she was, in many respects, his favorite), he had never, in her or her sister's recollection, discussed scientific matters with her. She remembers having been very taken with physics and mathematics in high school, but is aware of no mentors or heroes, male or female, who might have served as role models. She recalls only that in the middle of her junior year of college, at the end of a particularly exciting course (in genetics), her own interests in the subject were further encouraged by a special invitation from the professor to take the graduate course in genetics. From then on she was, unofficially, granted the status of a graduate student. She was given a cubbyhole; she could take courses. It would, however, be another year and a half before she could acquire enough credits to compensate for all the "Zs" she had accumulated earlier, and could officially graduate. By that time, there was no longer any question: "I knew I just had to go on." There was, however, a question as to exactly how, now that her graduate student status needed to be made official. "They didn't take any women in the plant breeding department, which is where genetics was given—at least they didn't want to take any graduate student women," she recalls. "They would take special ones, but not graduate students. But I was in botany, too, and I had taken a course in cytology which I'd enjoyed very much. (It was a course mainly in cell and chromosome work.) So I registered as a graduate student in the botany department, with my major in cytology (in chromosome work), and a minor in genetics and zoology." In addition to genetics and cytology (courses offered in the Ag school), she took a large number of courses in zoology. And what she didn't take, she audited.

• • •

The portrait that emerges from McClintock's recollections so far gives us only glimpses of the characteristics that would be so important in defining her as a scientist. As a child McClintock had a striking capacity for autonomy, self-determination, and total absorption. But what was truly exceptional was the extent to which she maintained her childlike capacity for absorption throughout her adult life.

A crucial component of this capacity was her wish to be "free of the body." She'd had a taste of this freedom—first as a young child "flying" along the beach, and, later, at moments of special concentration in her studies. "The body was something you dragged around," she says. "I always wished that I could be an objective observer, and not be what is known as 'me' to other people." Sometimes she managed even to forget her own name. She laughingly tells a story to illustrate how well she sometimes succeeded: "I remember when I was, I think, a junior in college, I was taking geology, and I just loved geology. Well, everybody had to take the final; there were no exemptions. I couldn't wait to take it. I loved the subject so much, that I knew they wouldn't ask me anything I couldn't answer. I just *knew* the course; I knew more than the course. So I couldn't wait to get into the final exam. They gave out these blue books, to write the exam in, and on the front page you put your own name. Well, I couldn't be bothered with putting my name down; I wanted to see those questions. I started writing right away—I was delighted, I just enjoyed it immensely. Everything was fine, but when I got to write my name down, I couldn't remember it. I couldn't remember to save me, and I waited there. I was much too embarrassed to ask anybody what my name was, because I knew they would think I was a screwball. I got more and more nervous, until finally (it took about twenty minutes) my name came to me. I think it had to do with the body being a nuisance. What was going on, what I saw, what I was thinking about, and what I enjoyed seeing and hearing was so much more important."

This capacity for total absorption, a wellspring of her creative imagination in science, took other forms as well. One of these was music. In college, she took a course in harmony, in which she had to write musical compositions, which the professor would play. " 'How did you ever think of that?' he would ask me. Well, I didn't tell him the reason I thought of that was that I had no other way of thinking—I hadn't had any experience."

What experience she accumulated in this course stood her in good stead when, in her senior year, she joined a jazz improvisation group playing tenor banjo at local places.

One time, she recalls with some amusement as well as wonder, when she was already a graduate student and very tired, she was playing at a dance: "I thought certainly I'd gone to sleep through the whole number, and at the end I woke up and asked the saxophonist, 'Did I fall asleep?' 'No,' he said, 'you were fine.' But I knew I was playing in my sleep; I was completely unconscious." By the end of her first year in graduate school, somewhat reluctantly, she gave up playing in the combo. "I couldn't be up late nights like that and still get enough sleep." From that point on, biology was to absorb virtually all of her passion.

CHAPTER 3

Becoming a Scientist

The men who taught Barbara McClintock in the early and mid-1920s, when she was a student at Cornell's College of Agriculture, were for the most part supportive and sympathetic. "One of the nice things about Cornell was that it was possible to get to know the professors. . . . Even outside classes, we would talk with one another." Lester Sharp, a cytology professor in the Botany Department, gave her a private course on Saturday mornings in cytological techniques. Later, he became her thesis advisor. She developed such skill that she became his first assistant, and by the time she was a graduate student, she could work on her own. "Sharp was not a research man himself," she recalls. "He had done some research earlier, but then he was mainly a person who wrote. He wrote a textbook on cytology—one of the first ones that came out—and he was excellent at reviewing the literature, but he was not involved in research." McClintock, however, was, as she put it, "research-minded," and Sharp gave her his full support. "He just left me free to do anything I wanted to do, just completely free."

This arrangement could not have suited McClintock better. By her second year of graduate school, she already knew what she wanted to do. The year before, while working as a paid assistant to another cytologist, she had discovered a way to identify maize chromosomes—to distinguish the individual members of the set of chromosomes within each cell—somewhat to the dismay of her employer, who had been working at the problem for a long time. "Well, I discovered a way in which he could do it, and I had it done within two or three days—the whole thing done, clear, sharp, and nice." He was not overjoyed with her success. "I never thought I was taking anything away from him; it didn't even occur to me. It was just exciting that here we could do it—we could tell one chromosome from another, and so easily! He had just looked at the wrong place, and I looked at another place." The end of a friendship, perhaps, but the beginnings of a career; having found the "right place to look," McClintock was to spend the years that followed doing just that.

Belling had just developed a new technique for cytological analysis that greatly simplified the preparation of slides for microscopic study of chromosomal structure. Technique, for cytological investigation, is the *sine qua non.* Everything depends on the care and ingenuity with which cells are fixed and stained on the slides. In Barbara McClintock's hands, Belling's carmine smear method underwent various modifications that made it particularly suitable for the study of maize and allowed the individual maize chromosomes to be observed throughout the course of their cycles of division and replication.

Chromosomes are found in the cell's nucleus. They first become visible as long slender threads in the early stages of nuclear division. Under suitable conditions, these "threads" can be then seen to double along the greater part of their length. Each chromosome thus consists of two half-chromosomes, or chromatids, that are held together at or near a region of the chromosome called the centromere—a structure that, later in the sequence, appears to regulate the movement of the entire chromosome. Throughout the first stage of division, prophase, the chromosomes become progressively shorter and thicker. Cells undergo two kinds of nuclear division—mitosis and meiosis—which, although they begin similarly, proceed in very different fashions.

In mitosis, the usual process by which cells duplicate themselves, prophase ends with the dissolution of the membrane that normally separates the nucleus from the cytoplasm (see the drawing below). In the next stage, metaphase, a spindlelike

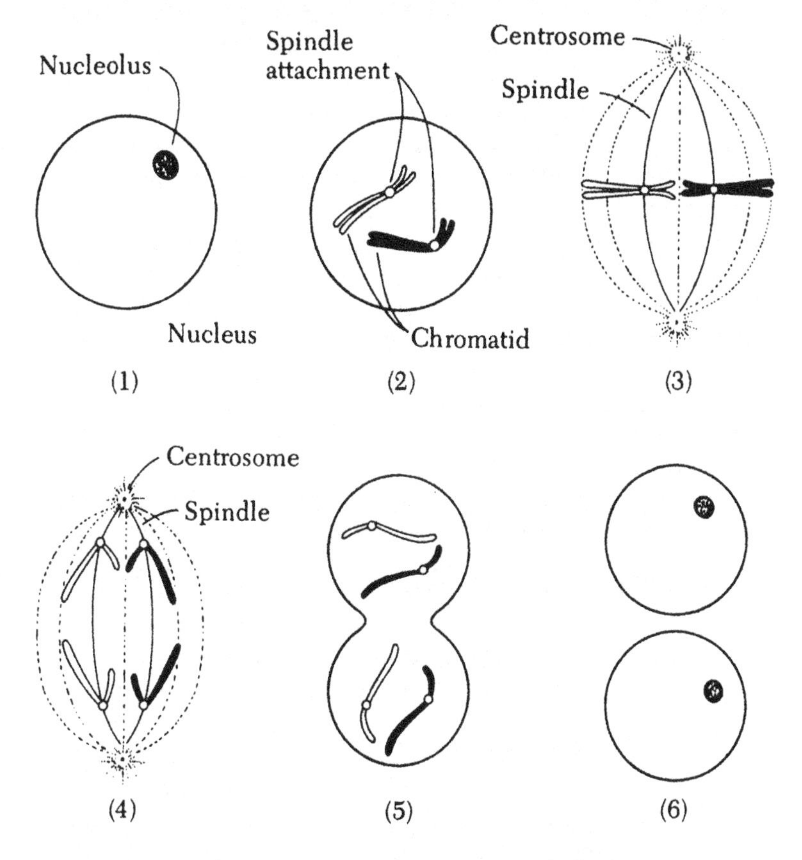

Mitosis in a nucleus showing only one pair of chromosomes (one chromosome black, the other white). There are usually many pairs in each nucleus. (Cytoplasm is not shown.) (1) Resting stage. (2) Prophase, each chromosome already double (consisting of two chromatids). (3) Metaphase. (4) Anaphase. (5) Telophase. (6) The two daughter nuclei in resting stage.

structure of fibers appears, radiating from opposite poles in the nucleus; the chromosomes attach themselves to these fibers, lining up in the middle of the spindle. In due course, the centromeres divide, thus completing the division of the chromosomes. In the following stage, anaphase, the two new chromosomes, led by the centromeres, move apart along the spindle toward opposite poles. Finally, in telophase, the nuclear membrane re-forms around each set of daughter chromosomes, resulting in two complete nuclei, each with the same number of chromosomes as in the original nucleus.

Sexually reproducing organisms require, in addition to a mechanism for duplicating normal cells, another mechanism for the production of gamete cells, which contain not a full double complement of chromosomes, but a single (haploid) complement. This mechanism is meiosis, and it differs from mitosis in several crucial respects. The full or diploid complement of chromosomes in a normal cell consists of two sets of homologous chromosomes, one set from each parent. In meiosis, each chromosome, instead of dividing, lines up side by side with its complementary chromosome (see the drawing on p. 43). This process of conjugation (or synapsis) occurs in the middle of prophase, in substages known as zygotene and pachytene. Later, the chromosomes in each pair pull a little way apart again (a substage called diplotene), but cross points (or chiasmata) can be seen at which pairs of chromosomes can exchange chromatids (crossing over). (It is above all the occurrence of crossing over in meiosis that makes this form of nuclear division of primary importance to geneticists.) At the end of prophase, a spindle appears, and the two complementary chromosomes proceed to separate along the fibers of the spindle, continuing as in mitosis, with one further exception. Following the first round of division is a second round of division in which, this time, the chromosomes do replicate themselves by longitudinal splitting. Four sets of nuclei result, each with a haploid set of chromosomes.

The miraculous coordination with which these processes are orchestrated makes the study of chromosomal dynamics of in-

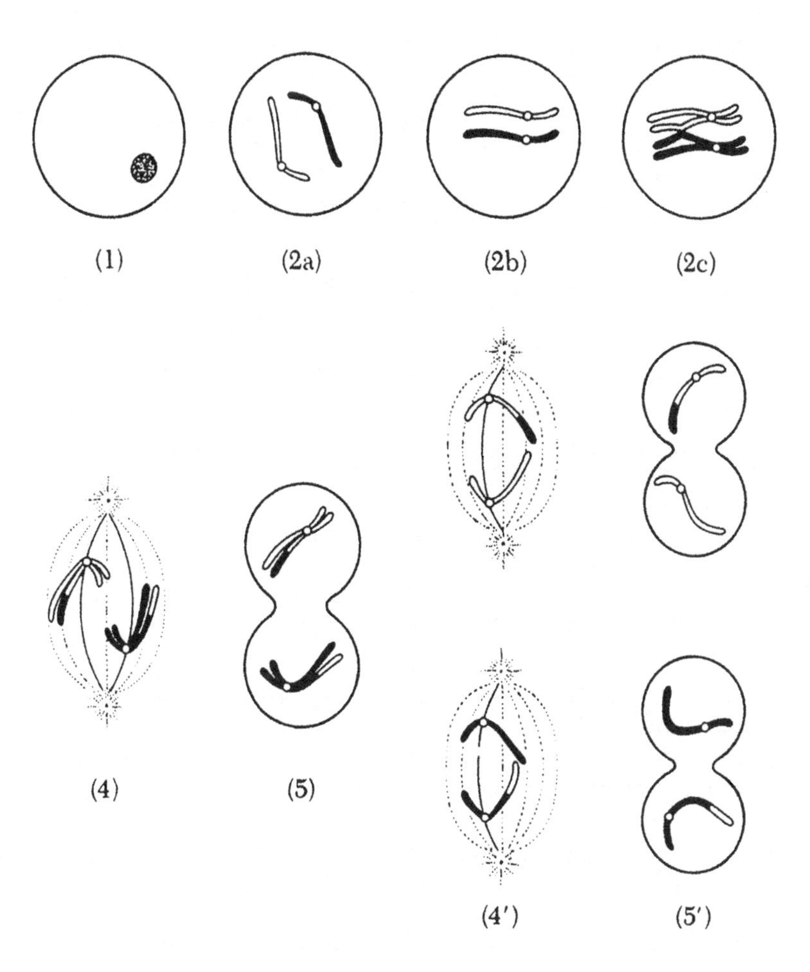

Meiosis. Stages are numbered to correspond to those of mitosis shown in the previous figure. Prophase in three stages: (2a) Appearance of chromosomes, (2b) pairing, (2c) separation of chromatids with chiasma formation. (4) and (5) First anaphase and telophase, two chromatids having mixed constitution due to chiasma. (4') and (5') Second anaphase and telophase. Four nuclei result from the whole process of meiosis.

terest in and of itself. But by the mid-1920s, when it was already widely accepted that chromosomes carried genes, the study of chromosomes had taken on special interest as a correlate to the study of mechanics of inheritance. However, the coordination of genetic and cytological investigations presented great problems. The two principal organisms for genetic study were *Drosophila* and maize. In *Drosophila,* while the individual chromosomes had been identified, they were so small that no fine structure had yet become legible, and in maize, the individual chromosomes had not yet been differentiated. They could be seen, and even counted, but they were simply "the chromosomes"—much as one might think of a brood of offspring as "the children." It was known only that maize had ten chromosomes (the haploid number). Now McClintock found that she could give to each one an identity. Each one acquired a label—a number, one through ten, ranging from the longest to the shortest—with which it could be followed throughout its life. Every chromosome, she found, had a distinctive morphology—a length, shape, and structure of its own. Some could be distinguished by the length of arms, by the pattern of beadlike structures (chromomeres) appearing along the chromosomes during prophase, and by the relative position of their centromeres. Certain morphological features varied considerably in different genetic stocks, suggesting the possibility of using these features as tags of particular genetic traits. In one strain in particular, a conspicuous, deep-staining knob could be found at the end of the second smallest chromosome (which she had labeled 9). These special features would become, in subsequent experiments, crucial landmarks in the exploration of unmapped genetic terrain.

As the geneticist Marcus Rhoades described the consequences of this discovery:

> Maize could now be used for detailed cytogenetic analysis of a kind heretofore impossible with any organism, and McClintock in the succeeding years published a series of remarkable papers

which clearly established her as the foremost investigator in cytogenetics.[1]

• • •

In 1927, when McClintock was not quite twenty-five years old, she completed her graduate work, received her Ph.D. in botany, and was appointed an instructor. By then, her next task was so clearly mapped out that the thought of leaving Cornell did not even arise: "What I wanted to do was something that is so obvious now it seems incredible that it was not obvious to the geneticists or plant breeders at Cornell at the time." It had already been established in *Drosophila* that "linkage groups" —sets of genes that are inherited together—were carried by specific chromosomes. "I wanted to do the same thing in maize —to take a particular linkage group and associate it with a particular chromosome. I had worked out the method, it was all going to go very well—there was no question about it. So I stayed on at Cornell to work on that."

The task was not an easy one, however, and McClintock needed help. "At that time there were two forms of geneticists: the breeder who did nothing but breeding, and the person who worked on chromosomes. They did not get together—they even worked in separate places." What McClintock was proposing to do was to combine them, and for this she sought the collaboration of someone who knew more about corn breeding than she did. "The people in genetics couldn't understand it. Not only that, they thought me a little mad for doing this. And the person who was going to work with me on this just kind of faded out of the picture." That may have slowed her down, but, "To me, it was so obvious that this was the thing to do that there was no way of stopping me." It meant that she would have to take over the other half of the research. "It was alright. I'm glad I did it."

To understand the disparity between Barbara McClintock's perceptions and those of the geneticists around her, a few historical markers may be helpful. McClintock herself explains:

"The subject of cytogenetics, that is, the relationship between the chromosomes and the genetic systems, was not clearly defined at that time. It had begun to be fairly well defined in *Drosophila,* but not in other organisms. Of course, later on, it was completely taken for granted." According to most historians, by 1927 the chromosomal basis of heredity had been accepted by almost all biologists for some years. It was by then more than a decade after the decisive experiments of Bridges and Sturtevant on *Drosophila* had provided direct cytological proof for the location of Mendelian factors on chromosomes. But even for *Drosophila,* cytogenetics was still a very young science. *Drosophila* chromosomes, as pointed out earlier, are extremely small, and only gross, relatively large-scale attributes could be seen under the microscope. The task of determining the location of specific genetic markers on individual chromosomes was proceeding steadily, but slowly. To some, it may have seemed a redundant exercise—a mere working out of details. To others, particularly to agricultural geneticists whose primary interest was, even at its most theoretical, in breeding, simply knowing that genes had a physical basis was enough. What Barbara McClintock perceived, as no doubt did the biologists working on the cytogenetics of *Drosophila,* was that in opening up the cytological window, as it were, one was embarking on the long pursuit of the physical character of the genetic process. A host of new questions of considerable theoretical importance could be asked, questions having to do with the relations of genes to one another, and with the mechanics of genetic variation. But the first task for McClintock was to locate the markers of known genetic traits of maize on the individual chromosomes. Although the corresponding task was well on its way for *Drosophila,* it had barely begun in plants. *Zea mays* was an obvious candidate for two reasons: first, more was known about the genetics of maize than of any other plant, and second, with McClintock's new techniques, maize chromosomes could now be seen with much greater detail than could *Drosophila* chromosomes. Later, when the enormous salivary chromosomes of *Drosophila* were discoverd by T. Painter in 1933, that would all be changed. But in the meantime, maize chromosomes permitted an analysis of fine structure far beyond what was possible on *Drosophila.*

About this time, "a person came to Cornell who had gotten a master's degree in genetics and was acquainted with the *Drosophila* work. He had come to get his Ph.D. in genetics at Cornell, to work on maize, and was going around asking people what they were doing. He came to my little table in the big laboratory, where a lot of other people were working as well, and asked me what I was doing, and I told him. Well, he became very excited and went around and explained to the others the significance of what I was trying to do. As a consequence, I was taken back into the fold."

This person was Marcus Rhoades, then a young graduate student, later a leading geneticist. His meeting with McClintock was a signal event in both their lives. It marked the beginning of a lifelong friendship from which two careers have continued to benefit. To McClintock, at that time, it provided a much needed intellectual companionship: "It was so obvious to him," she remembers. "He *understood* what I was trying to do when the others did not." To Rhoades, it gave access to a new field: "He had wanted to get into the area of cytogenetics, and he saw that here was an opportunity. Ever since, starting with the coincidence of his coming and speaking to me, he has been a cytogeneticist with maize."

George Beadle was another important figure in this period of McClintock's life. Years later, Beadle would gain international recognition for the pivotal role of his (and Edward Tatum's) "one gene–one enzyme" hypothesis in the development of molecular genetics. For the crucial experiment that bore out this hypothesis, he shared the 1958 Nobel Prize in Physiology and Medicine with Tatum and Joshua Lederberg. But at that time, George Beadle was a young graduate student, fresh from Nebraska where he had grown up in the cornfields. For all of these scientists, corn cytogenetics became a passion that would last their entire lives.

The magnet that drew these talented young men to Cornell

in the first place was Rollins A. Emerson, the foremost maize geneticist of his time. He was well loved by his students—as advisor, as Chairman of the Plant Breeding Department (not open to women), and as Dean of the Graduate School. Emerson created an atmosphere of hard work, enthusiasm, and openness that made his lab, in the words of several of his students, "a very special place" to be. But to Rhoades, at least, McClintock's work down the hall was even more exciting. From the first, he recognized that "she was something special."

While Emerson's research was confined to the study of corn genetics through breeding, McClintock's cytological work opened up a new frontier. The new kinds of questions that could now be studied drew McClintock, Rhoades, and Beadle together. Cytological proof of the basic postulates of genetics had thus far been provided only for *Drosophila*; now similar proof could be provided in plants. Furthermore, through cytological analysis, one could begin to map out the mechanical processes by which sequences of genes, located on chromosomes that could now be clearly seen, underwent variation from one generation to the next.

McClintock, recalls, "we were a group, all of us highly motivated, and we used to have our own seminars from which we'd exclude the professor—just us and a few others." These three were joined by Charles Burnham, Harold Perry, and H. W. (Pee Wee) Lee. Others, such as the brilliant geneticist-theoretician Lewis Stadler, were frequent visitors. In the view of most researchers in the field, this was the Golden Age of maize cytogenetics, beginning in 1928 and lasting as long as Rhoades and McClintock remained at Cornell. By 1935, it was essentially over. A photograph taken at one of their lunchtime seminars in the corn hollow—which on many occasions did in fact include Professor Emerson—survives as a historic document. It shows Barbara McClintock by the side of four tall and lanky, mostly young, men—a perky, attractive, and petite young woman who, but for an intense gleam in her eyes, might be described as "pleasant looking."

• • •

Maize cytogeneticists at Cornell University, Ithaca, New York, 1929. Standing, left to right: Charles Burnham, Marcus Rhoades, Rollins Emerson, and Barbara McClintock. Kneeling: George Beadle. (Courtesy of Marcus Rhoades.)

Today, Marcus Rhoades and George Beadle, both in their late seventies, are still growing corn—Rhoades at Indiana University in Bloomington, and Beadle at the University of Chicago. Tall and hardy, Marcus Rhoades looks at first sight more like a midwestern farmer than a renowned scientist. In truth, he is both. Delighted to talk about his old friend, and eager to show off the lush beauty of his campus, Rhoades reminisced expansively about old times. Every scientist comes to his subject with a world view that is uniquely his own—a world view reflected in his relations to people as well as to his subject. Each brings a distinct set of interests—interests stamped by his or her own personality. Rhoades's personality is marked by an open and generous warmth. Retired since 1974, he still never misses a day in the lab. He loves his work and can imagine no more pleasurable way to spend his retirement. There he is, seven low-keyed days a week. But he delights in the idiosyncracies of his friends and colleagues as much as he delights in the idiosyncracies of maize.

Recalling his first days at Cornell, he confirms McClintock's account of his role in mediating between her and the geneticists. "One thing that's to my credit—that I recognized from the start that she was good, that she was much better than I was, and I didn't resent it at all. I gave her full credit for it. Because —hell—it was so damn obvious: she was something special."[2] According to Rhoades, she was the real inspiration of their little group. "I loved Barbara—she was tremendous."

Over fifty years later, he sees no reason to revise his early judgment. "I've known a lot of famous scientists. But the only one I thought really was a genius was McClintock."

Most of McClintock's colleagues knew her to be very bright, but many found her somewhat difficult as well. Rhoades explains: "Barbara couldn't tolerate fools—she was so smart." And evidently, at least at times, she was impatient with those who could not keep up with her. Nor did everyone understand the significance of her work. Both factors, in Rhoades's view, contributed to her initial difficulty in communicating with geneticists. Certainly they accepted the chromosomal basis of inheritance, but "the advances hadn't been very great." The much higher level of precision and detail in McClintock's integration of cytology and genetics was new to them. "The students were aware of the meaning of her work, but some of the older folks weren't—their minds were closed."

George Beadle's recollections are less specific.[3] He is unequivocally a biologist in the modern tradition. More interested in the molecular mechanisms of genetic systems, he lost track of McClintock's work fairly early on. But he remembers her cytological prowess with great respect. He also remembers her unbridled enthusiasm. In a brief autobiographical piece he wrote for the festschrift on the occasion of Max Delbrück's sixtieth birthday in 1966, Beadle reminisces about his early cytological work on pollen sterility. He remarks: "My enthusiasm was shared—so much so in the case of Barbara McClintock that it was difficult to dissuade her from interpreting all my cytological preparations. Of course, she could do this much more effectively than I."[4] He still remembers the occasion that perforce had turned into a joint publication; McClintock had simply beaten him to the interpretation of his own data. He laughs now—reasonably confident that he could have done it himself—but at the time he was irked. Still, he remembers their lunches in the corn patch as "a high spot." Though he never felt personally close to McClintock ("she wasn't like the rest of us") and distances himself from what he calls her "mysticism," his respect is ungrudging. "She was so good!" He refers to her work as "fantastic," "spectacular," "the best job that's been done," even though it's not his "kind of stuff." In his retirement, as Emeritus President of the University of Chicago, Beadle has returned to the cornfields. He plans to devote his remaining years to tracking down the origin of maize, a subject on which McClintock and he evidently do not see eye to eye.

The old days in the cornfields at Cornell were particularly happy ones for McClintock—and prolific, too. Between 1929 and 1931, she published nine papers detailing her explorations of the morphology of maize chromosomes, and her successes in correlating the new cytological markers with known genetic markers. Each was a major contribution to the field. She remembers Rhoades and Beadle as more than making up for whatever support and stimulation she might otherwise have lacked. "We were opening up a new field, the three of us were." Their mood was confident: their work together was beginning to bring them recognition. But between McClintock and the other two, an important difference lay quiescent.

For Rhoades and Beadle, Cornell was but the first step along a well-marked path. Their careers would unfold along the lines expected of brilliant, hardworking, and ambitious young men.

For McClintock, the road ahead was uncharted. She was simply doing what she wanted to do, with "absolutely no thought of a career." Nor, it might be said, could she realistically have thought of a career. In Barbara McClintock's day, women in the sciences tended to be scientific workers and teachers rather than scientists, pursuing science more as an avocation than a vocation. Careers as research scientists were not available to them.[5] Positions in the universities that were open to women were for the most part limited to assistantships and, occasionally, instructorships. They might teach in the women's colleges, or they might marry scientists and work in their husbands' labs. For most of these young women, their love for science was sufficient reward; they adapted to their situation. Barbara McClintock, by contrast, could not, or would not, adapt to the limitations imposed on her sex here any more than anywhere else. She may not have thought in terms of a career, but neither did she think in terms of the alternatives that other women in science seemed to accept; she knew who she was and where she belonged. She was passionate about her research, and she was good at it.

Another woman from the Golden Age of maize cytogenetics —junior to Beadle, Rhoades, and McClintock, but a significant presence nonetheless—provides a valuable perspective. At the end of the summer of 1929, a twenty-year-old Wellesley graduate named Harriet Creighton arrived in the Botany Department at Cornell. By 1931, she would achieve worldwide recognition as coauthor with McClintock of the paper that provided the conclusive evidence for the chromosomal basis of genetics.

Half a century later, Dr. Creighton, now Professor Emeritus at Wellesley, is a robust and easy-mannered woman, relatively tall, more than half a foot taller than McClintock, with a strong, handsome face. She has the commanding air of someone who is confident of her place in the world, and who enjoys it. She speaks with a deep voice, made throaty by years of heavy smoking, and takes evident pleasure in telling a story. She seems especially delighted to talk about those early years.[6]

Having graduated from Wellesley the spring before, she had been steered to Cornell by Margaret Ferguson. Dr. Ferguson, an inspiration to many young women, had herself earned a doctorate from Cornell twenty-eight years earlier. In her many years of teaching at Wellesley, she was reputed to have trained more women botanists, and botanists' wives, than any other scientist anywhere. At Cornell alone, four of the botany professors were married to Wellesley graduates who all remained active scientists, working in their husbands' labs.

Wellesley graduated a large number of women who went on to graduate work in other fields of science as well; indeed, it ranks highest of all American colleges in the training of American women of science before 1920.[7] These women generally went on to institutions that were hospitable to the training of women scientists. In botany, they went primarily to Cornell or the University of Wisconsin. Harriet Creighton came to Cornell as a graduate student teaching assistant to Dr. Petrie, a paleobotanist, but, on her first day there, she was introduced to Barbara McClintock. Without formalities, McClintock asked, "What are you going to study?" Creighton didn't know—perhaps cytology, perhaps plant physiology. "Well," said McClintock, "I'd like to introduce you to Dr. Sharp; I think cytology and genetics would be better."

By the end of that first day, Harriet Creighton had her entire graduate program organized—all according to McClintock's recommendations. She would major in cytology and genetics, minor in plant physiology, and Sharp would be her advisor. But, for all practical purposes, she would henceforth be McClintock's charge. McClintock advised her as to what to study, where to live, when and what to avoid. "It was the best steering anyone could have given me," she recalls. Like the two other women who came to the department that year, Harriet Creighton's initial plan had been to enroll for a master's degree; she assumed that was in any case a necessary first step toward a doctorate. McClintock explained that it was not, and that, furthermore, she would be taken more seriously by the department if she were a doctoral candidate than if she were merely a master's candidate. Creighton was agreeable, ready to take McClintock's advice.

Had McClintock singled her out because she was a woman? "It could have had to do with being a woman," according to Harriet Creighton; "I don't know. We didn't think so much about it in those days, or at least didn't verbalize it." More to the point, she thought, was that McClintock had already begun to anticipate leaving Cornell and was looking for someone to groom as Sharp's next assistant. She recalled that at the time there was an active organization of graduate women in science —Sigma Delta Epsilon—and that McClintock had urged her to join, even though she herself seems not to have been a member.

Sigma Delta Epsilon, later to grow into a nationwide organization, began at Cornell in the early 1920s with about forty members at any one time. "Everyone joined." At first, its ambition was to provide a living center for the graduate women in science, but, by Harriet Creighton's time, they had abandoned the idea of maintaining a residential house. The group continued to serve as a kind of social and intellectual community, meeting now and then for dinners. At the very least, it provided an opportunity to get to know women from other disciplines. But for most of them, the principal focus of intellectual and social life was the lab, and Creighton learned more about being a woman in science from looking around her own laboratory than she did from the meetings of Sigma Delta Epsilon. She could see that Ezra Cornell's liberal vision of founding a university where "any student could receive instruction in any subject" did not extend to the faculty level. There, for example, was Miss Minns (Lua A. Minns, Department of Horticulture), who, though held in high regard, in her fifties was still an instructor. Seeing Miss Minns made an impression on Creighton that would loom large over the next few years as she sorted out what she would do.

In the meantime, the main business at hand was the learning of science itself. The subtle and difficult techniques of cytological analysis required a great deal of attention. But Creighton found that she was also learning a technique that may have been even more valuable; she was learning a method of following McClintock's discourse—which, even then, she recalls, many found dense and "hard to follow." Her discovery was that what sometimes seemed like non sequiturs in McClintock's line of thought were in fact a response "to the question you should have been asking at the moment"; she was addressing the doubt "you should have had." This discovery served Creighton well; it helped her, too, in building the confidence she felt was necessary to deal with the high standards McClintock imposed. "She was very quick to see things, and someone who wasn't quick had a hard time."

Toward the end of that year, in the spring of 1930, McClintock suggested a problem for Creighton to work on. She thought that it ought to be possible, by using the corn stocks that displayed the deep-staining knob she had observed on chromosome 9, to finally establish the correlation between genetic and chromosomal crossover that geneticists took for granted, but that had not been proven. (Genetic crossover is observed when an organism combines the traits of both its parents corresponding to two genes that are normally linked. "Linked" genes are normally inherited together and hence assumed to be on the same chromosome. It had been assumed by most geneticists that the physical basis of this event was an actual physical crossing over of segments of the paired chromosomes in question, resulting in the inheritance of a chromosome that derives partly from one parent and partly from the other.)

McClintock had already determined the location of a particular group of linked genes on the same chromosome and was in the midst of working out the morphology of additional cytological markers. All that was needed were two cytological markers on the same chromosome, located near two distinct genetic markers. By simultaneously following these two sets of markers through genetic crosses (matings) with plants that did not have these markers, it would be a simple matter to resolve whether or not both kinds of crossover take place concurrently. Creighton's initiation into the art of corn genetics began with the seeds (or kernels) that displayed the genetic and cytological markers that McClintock had isolated.

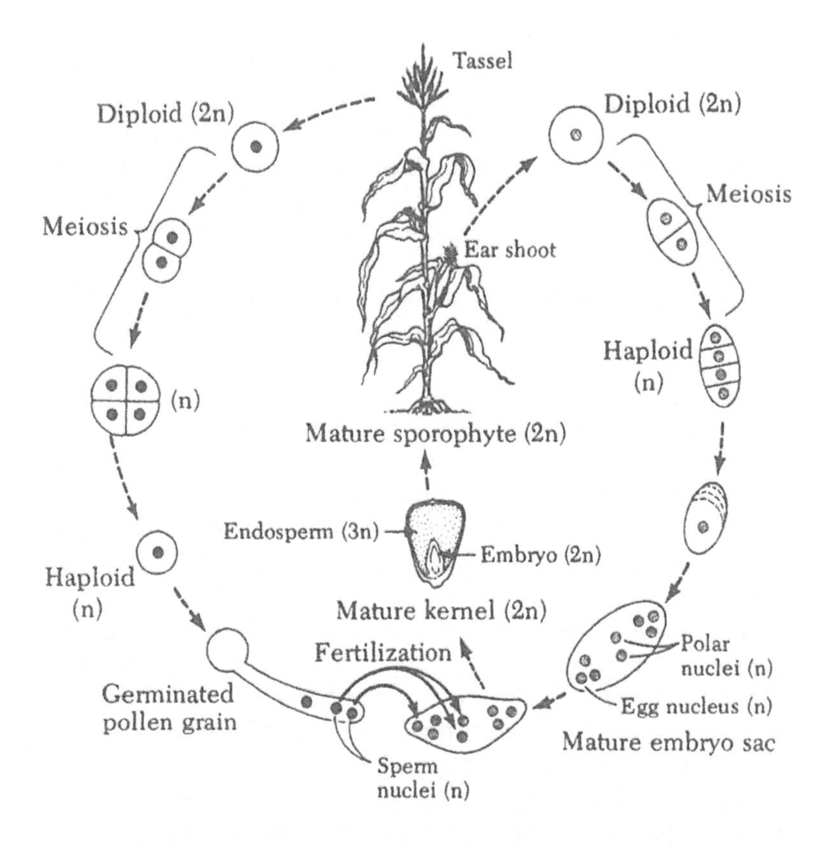

The life cycle of corn. Mating in corn occurs midway through the growing season, when the green plant is large and fully developed, but the ears just barely formed. The pollen matures in the tassel at the top of the stalk, while the embryo sacs are buried in the ears—one embryo sac in each future kernel. Pollination, normally subject to the vicissitudes of wind and air, takes place when the pollen grain falls from the tassel and settles on a silk extending out from the embryo sac to the outer end of the young ear. In order to control mating, extreme care must be taken to ensure that the only pollen that comes into contact with the female flower is pollen from the designated paternal plant. This is normally accomplished by tying a paper bag over the ear shoot before the silks appear. The pollen to be used in the cross is collected by enclosing the appropriate tassel in a different paper bag. Fertiliza-

Corn genetics is hard work. To prolong the growing season, it is necessary to plant the corn in the warmest spot available —usually a hollow facing south. As the summer progresses, the heat can become oppressive. Work begins early in the morning, before it gets too hot, but continues throughout the day. The young plants need constant watering; they must not be allowed to dry out. Each one is tagged and watched carefully, both in the fields and in the laboratory. When the time for fertilization arrives, utmost precaution must be taken to prevent a chance pollination.

After a long hard day of such work, one had earned one's rest. For relaxation, McClintock loved nothing better than a good tennis game. Every day at five o'clock, weather permitting, she would meet Creighton on the tennis court—"as intent on returning a ball as she was on hunting a chromosome." The younger and considerably taller woman, although herself an accomplished tennis player, recalls being run ragged.

Harriet Creighton describes Barbara McClintock's generosity in giving the seeds for this important project to such a novice as herself as fitting a tradition established by Emerson.

---

tion takes place when the pollen grain (containing two sperm nuclei) comes into contact with the silk, whereupon it sends out a long tube that grows down through the silk into the embryo sac. One sperm nucleus fuses with the egg nucleus, and the other (identical to the first) fuses with the two polar nuclei (genetically identical to the egg) in the embryo sac. This fused nucleus (containing three sets of chromosomes, 3n, one paternal and two maternal) gives rise, through subsequent mitotic divisions, to the nutritive tissue that surrounds the embryo and comprises the endosperm, the bulk of the mature kernel. Both embryo and endosperm carry the same genetic alleles; both are products of the same fertilization. The great advantage of corn genetics is that the endosperm matures with the parent plant, thereby providing the investigator with a preview of the genetic character of the plant that will grow from the embryo in the following season. Cytological examination can be performed both on the endosperm tissue and, later, on the tissues of the new plant.

It was his policy, she recalls, to give a new student "the best and most promising problem you have." Young Creighton herself hardly realized its significance and, by Rhoades's recollection, required constant prodding from McClintock to get it done. It was not until well into the summer that it dawned on her that no one had actually solved this problem before, that it was not just an educational exercise but a substantial contribution to the field. To this day, McClintock shrugs it off—"It was such an obvious thing to do"—but she also acknowledges that it needed to be done.

Without question, it was a timely piece of work. Across the Atlantic, Curt Stern's work with *Drosophila* was progressing to the point where he, too, could anticipate doing the experiment that would put "the final link in the chain" of classical genetics. Microscopic examination of the meiotic stages of *Drosophila* had, throughout the 1920s, continued to prove intractable. Finally, Stern was able to delineate cytological markers that were sufficiently clear to follow through the process of crossing over. The experiment that would also establish the correlation between genetic and chromosomal crossover in *Drosophila* was well under way. Stern would surely, according to Creighton, have beaten them to publication if T. H. Morgan had not intervened.

During the spring of 1931, Morgan came to Cornell to deliver the annual Messinger lectures. Once the lecture series was over, the great geneticist emerged from seclusion and made the rounds of the laboratories; he wanted to know what everyone was doing. When he came to the little office that Creighton and McClintock were then sharing, Creighton told Morgan of her project and showed him the preliminary results they had from the past summer's crop. Immediately, he asked if they had yet written this up for publication. No, they were waiting for the next crop of corn to confirm their initial data. Morgan demurred. He thought they had quite enough already; they should publish their results now. Overriding everyone's hesitations (Sharp, for example, pointed out that this was Creighton's Ph.D. dissertation, and she had three years of residency to fulfill), Morgan asked for pen and paper. On the spot, he wrote a letter to the editor of the *Proceedings of the National Academy of Sciences* telling him to expect the article in two weeks. The paper arrived on July 7 and appeared in August 1931.

Curt Stern, whose parallel work was by then well on its way, was scooped. His paper was based on more extensive data, but it did not appear for several more months. Stern was visibly perturbed. Late in his life, Stern reminisced about the day he publicly presented his work:

> I gave my paper with the enthusiasm of a successful youth. Soon after, one of my colleagues from the Kaiser Wilhelm Institute came to me and said: "I didn't want to spoil your fun but while you were on vacation a paper came out written by Harriet Creighton and Barbara McClintock who did experiments in maize equivalent to what you just announced as unique." May I confess . . . that I am still grateful to my colleague for permitting me the feeling of triumph for half an hour longer than I would have had it if he had told me about the Creighton-McClintock paper *before* my talk.[8]

According to Creighton, Morgan later confessed that he had known about Stern's work at the time. But, as he explained (about a year after his intervention), he was also aware of the fact that, even though Creighton and McClintock had begun the summer before, it would have been a simple matter for Stern to overtake them. With *Drosophila,* one need not wait an entire growing season to learn the results of genetic crosses; one can get a new generation every ten days. Creighton recalls Morgan's saying, "I thought it was about time that corn got a chance to beat *Drosophila!*"

All of the principal characters of this story had a chance to meet the following summer when the Sixth International Congress of Genetics convened in Ithaca, New York. It had been five years since the last occasion on which the Congress, with 836 members from thirty-six countries, had had the opportunity to meet at all, and thirty years since it had met in the United States. The meeting opened on August 24, 1932, with 536 geneticists registered (many of the European delegates had been unable to attend). All the luminaries of the field were present. T. H. Morgan was President, and Rollins Emerson Vice President. Richard B. Goldschmidt, Director of the Kaiser Wilhelm Institute in Berlin, acted as spokesman for the European contingent.

In his opening address, Morgan undertook to review the history of genetics and to assess the current state of the art. By way of summary, he concluded with a list of the five most important problems for geneticists in the immediate future. First was an understanding of "the physical and physiological processes involved in the growth of genes and their duplication"; second, "an interpretation in physical terms of the changes that take place during and after the conjugation of the chromosomes"; third, "the relation of genes to characters"; fourth, "the nature of the mutation process"; and fifth, "the application of genetics to horticulture and to animal husbandry."[9]

General papers were given in the morning. They focused primarily on the areas of Mendelianism, evolution, chromosome mechanics, and mutation. In one of these, Emerson reported on the current status of maize genetics; in this he referred extensively to the work of McClintock on the determination of the chromosomal location of known genetic linkage groups. Other aspects of her cytological investigations were reported in papers given by H. J. Muller and Lewis Stadler, and the Creighton and McClintock work received special mention in the papers of Karl Sax and (of course) Curt Stern. Stern had been invited to review the genetics and cytology of crossing over (which he did in German). The afternoons were devoted to sectional meetings, running five or six at a time, and covering a somewhat broader range of issues. McClintock delivered a paper on the occasional pairing of nonhomologous parts of chromosomes—a subject on which she would subsequently elaborate further—and served as Vice Chairman of another of the afternoon sections. In addition, she and Creighton prepared an exhibit illustrating their cytological evidence for four-strand crossover.

A group photograph taken at the meeting shows McClintock and Creighton in the second row on the far right—two out of seventy women in a group of 389 attendees. Her colleagues of that time describe this as probably the high point of Harriet Creighton's career as a research scientist. Two years later she would leave Cornell for a teaching job at a women's college. For Barbara McClintock it was only the beginning.

After the Congress was over, a chance and especially felicitous encounter occurred on a transatlantic steamship. Rhoades remembers McClintock's great delight and pleasure when she heard about it. It seems that Dr. and Mrs. McClintock were embarked for a holiday in Europe when they struck up a conversation with one F. A. E. Crew, a Scottish geneticist who was on his way home from the Congress. When it emerged that these were Barbara McClintock's parents, Crew had the opportunity of being the first to inform them of their daughter's great scientific success. In a moment, years of misgiving, disapproval, and worry about their youngest daughter's "odd" choices gave way to pride.

CHAPTER 4

A Career for Women

In 1931, a year before the Sixth International Congress of Genetics, McClintock had come to feel that, finally, it was time to leave Cornell. "I couldn't stay indefinitely—there was no job for me. I don't know that they'd even have wanted me to stay on as an instructor. It would have been embarrassing for them and embarrassing for me." Although she had abundantly proved her worth and had won the respect, support, and affection of many of her colleagues at Cornell, a proper faculty appointment was out of the question. Not until 1947 did Cornell appoint its first woman assistant professor in a field other than home economics. Like most women of her generation, McClintock knew the score, and, as long as she was able to continue her research, she seemed to accept it. Time and a place to work were all she claimed to need.

Fortunately, she was awarded a fellowship from the National Research Council, which gave her support for the next two years. Between 1931 and 1933, she divided her time between the University of Missouri, California Institute of Technology, and Cornell. Cornell continued to serve as home base—a stopover on the route between Columbia, Missouri, and Pasadena, California, or wherever else her travels would take her. She retained her place in the lab at Cornell, returning to check her crops, and borrowing equipment as needed. "They couldn't supply me with a job, but they supplied me with everything else; they couldn't have been nicer," she said.

In more senses than one, Ithaca had long since become a second home, replacing even her first in importance. This was the place to return to. During one time when she was sick—shortly after college—her physician, Dr. Esther Parker, invited her to her own home to convalesce. Dr. Parker lived with her mother and was accustomed to having students come and stay when they needed to be taken care of. "So I went, and stayed longer than one normally would. We just got along very well—I got along well with the mother and the dogs and the canaries—it just became a good place for me to be. I liked working in the garden, cutting the grass, answering the phone, and so forth. It was a relaxed, easy, pleasant experience—another example of my good luck. I had good luck at Cornell all the way along the line."

Esther Parker became a close personal friend. The friendship continued for many years as an emotional and physical anchor in McClintock's life. Dr. Parker's household provided another family and "a home to go to. . . . It was a home for me where I could really keep in touch and be involved with my work; I wasn't stranded in the sense of just going here and going there. . . . I wasn't lost." With a place to live and a place to work, Cornell became her established headquarters. "That was where I wanted to be when I had no job—I was always there."

But in 1931, she began to explore the outside world. That summer, an invitation from Lewis Stadler brought her to the University of Missouri. Stadler had come to Cornell back in 1926 on a National Research Council Fellowship to work with Emerson. During that time, he and McClintock had become close friends and colleagues, beginning a collaboration that was to last for many years.

The two shared a keen interest in the genetic composition of maize, and Stadler's investigation of the mutagenic effects of X rays—discovered independently by H. J. Muller in 1927—greatly excited McClintock's imagination. To this day, mutations are the mainstay of genetic research, and in the early days progress was limited by the need to rely on their spontaneous occurrence. By vastly increasing the frequency and variety of mutations, X rays could greatly expedite the probing of genetic structure. McClintock was eager to participate in the new research.

The technique was to irradiate the pollen grains of plants carrying dominant genes of particular traits, and then to use the irradiated pollen to fertilize kernels of plants carrying recessive genes for the same traits. It emerged that the X rays were inducing large-scale changes in chromosomal arrangement—changes that would show up in a wide range of visible alterations in the young plant, most dramatically in the coloring and texture of the kernels. McClintock's challenge that summer was to identify the specific nature of these chromosomal changes. The same cytological techniques she had earlier developed now enabled her to determine the minute physical changes within the chromosomes that were induced by the X rays. She found translocations, inversions, and deletions of parts of chromosomes, all resulting from exchanges between normal and damaged chromosomes occurring in the course of meiotic division. "That was a profitable summer for me! I was very excited about what I was seeing, because many of these were new things. It was also helping to place different genes on different chromosomes—it was a very fast way to to it."

Her story of the identification of one of these new phenomena illustrates McClintock's unique style of thinking. "As I was going through the field, I saw plants that were variegated—part dominant and part recessive. I didn't look at the variegated plants, but somehow or other they stuck in my mind." That fall she received a reprint from California in which variegation was described. A small chromosome was also identified that might, it was suggested, be related to a genetic fragment that seemed to be "getting lost," thereby causing the pattern of variegation. "As I was reading this paper I said, 'Oh, this is a ring chromosome, because a ring chromosome would do this.' " Although a ring chromosome had in fact recently been observed for the first time, neither McClintock nor her colleagues yet knew of its existence. Her reasoning was as follows: "Everything had been discovered that you would expect if you broke a chromosome and the fusions occurred two by two. That is, if you had a chromosome that had two parts to it, you could invert [one] and get an inversion. Or you could get a deletion. At this point you say, well, the reverse of a deletion is a ring chromosome. Why weren't people reporting ring chromosomes? They weren't. Therefore, these ring chromosomes must have a mechanism of getting lost. . . . So I wrote to these people and said, 'I think that what you have is a ring chromosome, and I think that they have a means of getting lost in sister strands exchange.' The response from California was: 'It sounds crazy, but it's the best thing we've heard.' "

The observed variegation patterns implied the occasional loss of certain genes, and the formation of ring chromosomes could account for this loss. Should a fragment break off from a chromosome, the ends of this fragment can anneal (or join) to each other. The ring chromosome thus formed can no longer participate in the normal mechanism of duplication and assortment; it will "get lost." The result is a deletion, the counterpart of a ring chromosome; the original chromosome will have reformed without the missing fragment.

So convinced was she that the variegated plants she'd seen the summer before had ring chromosomes, that she wrote to Stadler asking him to grow more of the same material so that she could examine them and find out. Stadler was glad to do so, and she went, arriving in Missouri approximately two weeks before the plants were ready for cytological examination. "I was so full of ring chromosomes they began to kid me. They didn't take it seriously, and they kidded me for a couple of days, but when we went out to the field and looked over the top corn, the first thing I noticed was that they were calling these plants ring chromosome plants. Then I got scared. I said, 'My goodness; they are now calling them ring chromosome plants when they have never seen a ring and don't know!' When the first plant was ready, my hand was actually shaking when I opened up the plant to get out the material to be examined. I took it right back to the lab and examined it immediately. It had rings, and it was doing exactly as it was supposed to do, and every other plant that I had deduced was a ring chromosome plant was doing exactly as I expected it to."

She was excited, of course, but also relieved. The thought that she might have convinced people by sheer enthusiasm disturbed her greatly. But, she asked, "Why was I *so* sure they were ring chromosomes? I could convince others to call them ring chromosomes before anybody had seen them. That was, God knows, true confidence. I was not trying to persuade anybody, but I was convinced. Why was I *so* convinced that the thing *had* to be, that it *couldn't* be anything else but that?"

And why did some others think it was crazy when she thought it was "absolutely legitimate?" Well, she thought it was legitimate because it was logical. "The logic was compelling. The logic made itself, the logic was it. What's compelling in these cases is that the problem is sharp and clear. The problem is not something that's ordinary, but it fits into the whole picture, and you begin to look at it as a whole. . . . It isn't just a stage of this, or that. It's what goes on in the whole cycle. So you get a feeling for the whole situation of which this is [only] a component part."

The next winter, when she was out in California, she was invited up to Berkeley and taken to the lab that had originally reported the fragment. "They had a lot of experts there, and they asked me would I look through the microscope, they didn't say anything [else]. I looked down, and what was there but a ring chromosome!"

Along with translocations, inversions, deletions, and ring chromosomes, another discovery—regarded by McClintock as one of her most important—was in the making. Clearly, she was well on her way to becoming a virtuoso at reading the intricate secrets of maize genetics. The previous year, she had gone out to spend the winter (1931–1932) at California Institute of Technology. T. H. Morgan had moved to Cal Tech in 1928 and was rapidly setting up one of the most exciting genetics laboratories in the country. George Beadle was one of his new recruits; after finishing his degree at Cornell, he had gone to Pasadena as a postdoctoral fellow in 1931. With Morgan, Beadle, and Morgan's wife, Lillian, McClintock had good friends to visit. And with an invitation to stay at the Morgan's home, she seemed settled for the winter.

The problem that caught her fancy was a small body normally visible at the end of chromosome 6, immediately adjacent to the point of the chromosome's attachment to the nucleolus—itself a relatively large round mass whose function was at the time not yet known. (It has since been established that the nucleolus is involved in the synthesis of ribosomes, which, in turn, are the "factories" of protein synthesis.) She had been observing this minute body for a long time, although no one else seemed to have noticed it. It was always associated with the nucleolus, and she felt certain that it must have something to do with the development of the nucleolus. In a particular case, which she noticed when invited to look at some material at Cal Tech, this body was split into two parts—one part in its normal position and the other associated with a different chromosome. Here might be an opportunity to investigate the function of this curious structure. "I don't know why, but I was sure it had the key."

Since no one else there wanted to pursue it, she arranged with E. G. Anderson, director of the lab, to come back the next winter in order to follow it up herself. Anderson grew the plants for her, so that when she returned a year later, everything was ready for her microscopic investigations. "It was . . . fascinating; I found that this thing organized the materials that were already there to make the nucleolus. That's why I called it the nucleolar organizer. . . . It takes the materials that are put into the chromosomes as the nucleolus disappears, in late prophase, and [that] then come out of the chromosomes and pass through that organizer, somehow, to make a nucleolus. . . . [The material] goes into the chromosomes and comes out and is reused. Without that organizer you do not get an organized nucleolus."

Given the state of knowledge at that time—a good three decades before biologists would be able to match a molecular description to the processes McClintock was observing under the microscope—her interpretation emerges as something of a tour de force. Marcus Rhoades recalls once saying to her: "I've often marveled that you can look at a cell under the microscope and can see so much!" She said, "Well, you know, when I look at a cell, I get down in that cell and look around." He laughs, "I'll never forget that," he says.[1] This is one of the many instances in which "looking around" paid off; without being able to say quite what it was she was seeing, she was able to arrive at a functional description that, except for the absence of biochemical terminology, accords remarkably well with contemporary analyses. Her principal conclusion stands: the nucleolar organizer region (NOR) must be present for a proper nucleolus to be formed. Her paper on this work is generally regarded as a classic,[2] but she feels that the organizational function of the NOR was, for the most part, not registered by the biological community. "I find that it was only a relatively few people. . . . who really got the point of the organization—of why I called it an organizer." Even today, almost fifty years later, she believes that "exactly what that means has still not really penetrated into cell biology. It is my feeling that it is much bigger than has generally been realized. They haven't [yet] recognized the meaning of organization."

In her original paper, McClintock was not able to spell out just how the NOR functions as an organizer, but she focused on a question that has been largely bypassed in the intervening efforts to analyze its molecular structure—namely, the question of function. Since that time, sophisticated techniques of analysis have shown that the NOR contains repeated sequences of DNA coding for ribosomal RNA, but her question of how it serves to "organize" the nucleolus has remained largely unaddressed.

A report of the Sixth European Nucleolar Workshop held in 1979 suggests that the topic is still very much a live one. It begins: "The workshop opened with an outline of our present understanding of nucleolar structure by M. Bouteille in which he emphasized the importance of the identification of the discrete pale-staining islands of the nucleolus-organizing regions."[3] The same report goes on to cite recent evidence establishing that "no organization of nucleolar material into nucleoli occurs in the absence of such a pale-staining 'fibrillar centre,' hereby shown to be functionally related to the nucleolus organizer." The actual organization of the nucleolus and its relation to the nucleolar organizer are, the author concludes, "still questions for the future."

McClintock herself takes responsibility for the fact that this early work was not understood; the paper in which it was reported was "too diffuse," and "very, very badly written." It was written in 1933, the year she went to Germany.

The two years of her National Research Council Fellowship were over. It had been a free, independent, and productive time in which she did what she wanted and went where she wanted to go. She had purchased a Model A Ford roadster to drive back and forth across the country, from Columbia, Missouri, to Ithaca, to Pasadena, and back. Still she had no thought of a career: "I was just so interested in what I was doing I could hardly wait to get up in the morning and get at it. One of my friends, a geneticist, said I was a child, because only children can't wait to get up in the morning to get at what they want to do." She tells a story about driving back from Cal Tech to Missouri. It was a time when news of a number of automobile accidents was fresh in people's minds, and she had been cautioned about driving. "My only concern was that if I were killed I'd never get the answer to that problem!" she remembers. Foremost in her mind was "purely the subject matter. I don't remember having any [professional] aspirations." Later, when she was in her mid-thirties, she remembers waking up and saying, "Oh, my goodness, this is what they call a career for women!" But that was still a couple of years off.

• • •

In 1933, recommended by an unmatchable trio—Morgan, Emerson, and Stadler—she received a Guggenheim Fellowship to go to Germany. Her original plan had been to work with Curt Stern, but Stern had by then already managed to leave Germany. Another great geneticist (also Jewish, but by virtue of his prominence, somewhat more secure than Stern) was still in Berlin, however. This was Richard B. Goldschmidt (1878–1958), head of the Kaiser Wilhelm Institute. Goldschmidt was one of the more colorful and controversial figures in the history of genetics. A thinker of wide scope who maintained an unflagging interest in the relation between genetics and development, physiology, and evolution, he was an outspoken critic of much of contemporary genetic theory for its excessively narrow and mechanistic focus; he particularly delighted in exposing the conceptual inadequacies of the American geneticists from the Morgan school. Although he later fell into disrepute among modern geneticists, at the time he was a highly esteemed figure. Many of his theories were ultimately discredited, but some of the ideas found a sympathetic echo in McClintock's own later views of genetic organization.

Nonetheless, 1933 was a grim time to go to Germany. "It was a very, very traumatic experience. I was just unprepared for what I [found]." Her political naivete at that time was shared by many Americans. "Had I been politically knowledgeable, I would have taken it in an entirely different way; I would have not been crushed and disturbed and utterly panicked by what I saw going on there." Perhaps, too, had she known more, she might not have gone. The tragedy of the Hitler regime hit her very hard, and apparently very personally. She was alone and helpless in the face of what was happening to those around her. Many of her close friends and associates dating back to her undergraduate days were Jewish, and whatever feelings about being an outsider had led her to seek out Jewish friends in college must have surfaced again in Germany. She doesn't like to talk about that period, which was the same one in which she wrote the nucleolus paper. "And that's why it's so bad. . . . I was in a bad state of mind."

Harriet Creighton recalls receiving frequent (sometimes daily) letters from Germany, letters describing cold, rainy, lonely days, no one to talk to, unable to get on with her work, and terribly lonely. Just before Christmas, she showed up unexpectedly, severely shaken, at her old lab at Cornell. This was where home was.

Back at Cornell, she picked up work as usual, but her depression was visible. On the heels of her traumatic experience in Germany, she came home to a nation reeling from the impact of the Great Crash. With an awareness sharpened by her own private experiences, she was obliged to face the realities of a situation guaranteed from the start to be difficult, but now vastly compounded by the economic depression around her. Ever since childhood, she had known she had to pay for going her own way; the price had seemed both necessary and manageable. Now, suddenly, the costs loomed larger than any she'd anticipated.

Where next? She had run out of fellowships and still had no regular position. Harriet Creighton learned an important lesson from watching McClintock—a lesson reinforcing the one she had earlier learned from watching Miss Minns: "You don't stay around Cornell, or any other university!"[4] Here was McClintock, seven years ahead of her in age, and more in experience, with the best of credentials, backed by the giants of the field, and unable to get a job. Creighton took a teaching position at Connecticut College for Women in 1934, where the pay was good and where she was sure to be appreciated. But McClintock was adamant. She was first and foremost a research scientist.

It must have been somewhere during this time that the recognition dawned on her that "this was what they called a career for women. It came very suddenly—it was startling and a little bit unpleasant. . . . At this stage, in the mid-thirties, a career for women did not receive very much approbation. You were stigmatizing yourself by being a spinster and a career woman, especially in science. And I suddenly realized that I had gotten myself into this position without recognizing that that was where I was going."

In the spring of 1934, seven years out of graduate school, with a worldwide reputation, McClintock hung on at Cornell without visible means of support. Times were hard for everyone, and university job opportunities were almost nonexistent, even for her male colleagues. Harriet Creighton had been fortunate, but her choice had removed her from the mainstream of biological research. Rhoades and Beadle, like many others in the mid-1930s, were biding their time as research associates, Beadle supported by Morgan, and Rhoades by Emerson. But McClintock had an edge over them of four and five years (respectively) of postdoctoral experience—with no prospects of support by anyone. She decided she had to leave; neither she nor her research could go on without a stipend.

Luckily, Emerson got wind of her decision and communicated with T. H. Morgan who, in turn, appealed to the Rockefeller Foundation—the one source from which support, at least for genetics, was still forthcoming. He requested $1800 to $2000 a year to finance McClintock's research in Emerson's lab. In an interview with Warren Weaver, then Director of the Natural Sciences Division of the Rockefeller Foundation, Morgan argued that such an investment "would be a most important contribution to the whole field of genetics. She is highly specialized, her genius being restricted to the cytology of maize genetics, but she is definitely the best person in the world in this narrow category."[5] Morgan also spoke of what he called her "personality difficulties," claiming that "she is sore at the world because of her conviction that she would have a much freer scientific opportunity if she were a man."

Emerson repeated the request in late June with a somewhat fuller discussion of Barbara McClintock's circumstances, also recorded in Warren Weaver's diary. Weaver's summary of Emerson's comments adds noteworthy detail:

> The Department of Botany does not wish to reappoint her, chiefly because they realize that her interest is entirely in research and that she will leave Ithaca as soon as she can obtain suitable employment elsewhere; and partly because she is not entirely successful as a teacher of undergraduate work. The

> Botany Department obviously prefers a less gifted person who will be content to accept a large amount of routine duty. At present McClintock has absolutely nothing in sight for next year. [Emerson expressed his fear that] "this situation will cause [her] so much worry that it may seriously interfere with her scientific work for a considerable period. She is admittedly nervous and high-strung, and very actively resents the fact that she is not given scientific opportunities. She feels that this is largely on account of her sex, since she has brains enough to realize that she is much more able than most of the men with whom she comes in contact."[6]

Noting that both Morgan and Stern concurred with Emerson that "it would be a scientific tragedy if her work did not go forward"—Emerson described her as "the best trained and most able person in this country on the cytology of maize genetics"—Weaver agreed that Emerson should put in a request for a research grant in genetics as part of his general program, to be quietly used to pay McClintock's salary for a year. The request was approved, effective October 1, 1934. The following summer, the Foundation renewed the arrangement for a second year, with the explicit proviso that this was to be the final year.[7] But, with no visible prospects for the future, the disparity between her own prospects and those of her male colleagues rankled. Appropriate positions may have been scarce, but they were being found for others. Rhoades says that her accomplishments merited more recognition than they received. "She'd earned it, and wasn't getting it."

Having made an initial investment in McClintock's career, the Rockefeller Foundation kept close track of the various efforts her friends and supporters made to locate a permanent position. Throughout 1935 Stadler was working on an appointment for her at the University of Missouri; C. W. Metz at Johns Hopkins[8] and E. W. Lindstrom wished to appoint her at Iowa State, but, as Hanson notes, "the Director of the Station will not appoint a woman."[9] There is even a report of an unofficial visit (to her great mortification) from her father. Dr. McClintock (who was by then working for Standard Oil as an industrial surgeon) sought out Hanson to solicit the assistance of the Rockefeller Foundation in securing a permanent post for her.[10] Of course, even though the foundation did operate as an unofficial clearinghouse for evaluations and prospects of young geneticists, such direct intervention was not its policy.

McClintock found an opportunity to apologize for her father's embarrassing and uninvited visit a couple of weeks later at the Genetics Meeting in Woods Hole. Hanson's informal notes report the conversation that ensued. He explains that he and Warren Weaver

> were very happy to meet her father and that if it will be of any comfort to her to know it, the decision to renew her grant was made before her father's visit and in any event the visit could not have influenced our decision one way or the other. This apparently leads [sic] her to go on and tell something of her early history, which may account in part for the kind of person she is today. Miss McClintock has a slight, boyish figure, weighing about 90 lbs., with a tousled boy's haircut. She said she was the youngest of 3 girls and her father was so greatly disappointed that she was not a boy that he proceeded to raise her as a boy. He got her boxing gloves when she was 4 years of age and as she came along she was provided with boys' toys and played boys' games. And she still looks and acts more boy than girl. She says the $1,800 of her present grant is the largest income she has ever had but that she is not interested in money beyond enough to live on. In her work at Cornell she has to have a car as she travels 100 miles every three days going back and forth to the experimental plots. She has to pay the expenses of the car out of her grant. Says she has bought no clothes for years and looks it.[11]

In the same conversation, McClintock makes a point of explaining "that her assistantship to Emerson is purely a fiction, as he has allowed her to work entirely on her own problems." Proud as always, she wanted to set the record straight. She was her own person, and did her own work, needing little from the outside world.

But Hanson's emphasis on her boyishness may have missed the point of her own remarks. Although adamantly rejecting female conventions, her wish was not to be "more boy than girl," but to transcend gender altogether. Given enough time person-to-person, it seemed—in her own eyes as well as in those of others—that she succeeded. "When a person gets to know you well, they forget that you're a woman. . . . The matter of gender drops away." She laughs as she recalls a conversation from a later period when she'd become a professor, in which a young graduate student volunteered: " 'I can't stand women professors, I just can't stand them.' Well, I just let him go on a while, and then I said, 'Herschel, to whom do you think you're talking?' "

But at this point in her life there is a certain irony in her assertion of freedom from sexual stereotypes. More poignantly than ever before, she was learning of the many ways in which, in the real world, the matter of gender does not drop away. Now, she says, "outside, it's always there, always intruding." No efforts of her own would erase the fact that she was a woman in a profession institutionally established for men. And no assertions of her own self-sufficiency could alter her very real dependence on the world around her. Even if she needed nothing else, she needed a job.

Without question, McClintock posed a severe problem for her colleagues. Fair-minded men, they readily acknowledged her merit. They recognized fully just how able she was and how important her contributions were to the growth of genetics, and they were willing to go out of their way to help her find support. As individuals, they did not hold the fact that she was a woman against her. The difficulty was that jobs for women were few. And compounding that difficulty was McClintock's own attitude. In effect, she was refusing to accept a woman's place. She would not be a "lady scientist" any more than she would be a "lady" in a more conventional domain. She had come to insist on her right to be evaluated by the very same standards as her male colleagues. Instead of being grateful for the efforts made on her behalf and the rewards she did receive, she resented the palpable disparity between her opportunities and those of her colleagues. She insisted on equating merit with rights.

This made her not only anomalous, but, in the eyes of many of her colleagues, problematic; she was seen as having "a chip on her shoulder." Morgan interpreted her "conviction" that she would have a much freer scientific opportunity if she were a man as evidence of her "personality difficulties." And not many scientists want a colleague whom they find difficult. Even for her most devoted supporters, finding a professional niche for her was no easy matter.

For the next two years, with the support of the Rockefeller Foundation, McClintock continued to pursue her own research in Emerson's lab, but her spirits were low. She published two papers in 1935: one a corroboration of her earlier work with Harriet Creighton, and the other a review of recent progress in maize genetics, which she wrote with Marcus Rhoades. In *A Short History of Genetics,* L. C. Dunn writes: "The latter publication marked the highest point attained up until that time in unifying cytological and genetic methods into a single clearly marked field."[12] It also marked the end of the Golden Age of maize cytogenetics at Cornell. By the end of 1935, McClintock's friends from that glorious period were widely scattered. Rhoades was with the U.S. Department of Agriculture; Beadle, having been at Cal Tech since 1931, was now on his way to Harvard; Harriet Creighton was at Connecticut College for Women. Only McClintock remained at Cornell, and, for the first time since she began her career, a gap appeared in her list of publications. She published nothing in 1936.

CHAPTER 5

1936–1941: University of Missouri

Among Barbara McClintock's supporters, Lewis Stadler was perhaps the most energetic in his efforts to find a position commensurate with her abilities as a scientist. A native of Missouri, Stadler had been at the University of Missouri in Columbia since 1919 and on the faculty since 1921. He pushed hard for an appointment for McClintock.

In the mid-1930s, with an $80,000 grant from the Rockefeller Foundation, Stadler was in the process of building a major center of genetics at Missouri and was especially eager to have McClintock as a colleague. He was finally able to persuade the administration to offer her a position as assistant professor in the spring of 1936. In many ways, it was not an unattractive offer. The salary was not much of an improvement over her fellowship,[1] and the rank hardly commensurate with either her scientific maturity or reputation, but it was her first offer of a faculty position; it provided her with a laboratory, a modicum of security, and a chance to get on with her research. She accepted the position and, with considerable relief, settled in and resumed the work that, more and more, had become her daily sustenance. The task of tracking the dynamics of chromosomal variation continued to absorb and reward her. McClintock had helped to blaze a trail that researchers around the country now pursued. But for sheer perspicacity, she remained unequalled. No one else could learn quite so many of the cell's secrets simply by close observation. Using the two kinds of evidence available to the cytogeneticists—the one coming from the wealth of new patterns of color and texture in the tissues of the maturing plant that could be seen with the naked eye, and the other from the physical changes in the chromosomes, observable only through the microscope—she had developed a unique virtuosity at integrating these disparate clues into a coherent and meaningful whole. Her ability to identify those clues that were worth following, her instinct for what was important, grew steadily. Marcus Rhoades has described her as having a "green thumb": "Everything she did would turn into something big!"[2]

One of the things she began during this time at Missouri turned into something very big indeed—partly in itself, and partly because of the possibilities it introduced in her subsequent work. After completing her work on ring chromosomes, she turned to an investigation of the manner in which broken chromosomes tend to reanneal. X rays characteristically induce many chromosome breaks, producing fragments that tend to reanneal in either normal or inverted sequence. She found that a chromosome with certain kinds of inversions can, through crossover with a normal homologous chromosome, produce a "dicentric" chromosome, that is, a chromosome having two centromeres, or cell division poles. In each subsequent round of nuclear division, the sister halves of such a chromosome attempt to separate (during anaphase), but remain bridged by the chromatin between the two poles. As the mechanical stress mounts, the bridge breaks, and, when the chromosomes reduplicate, the new pairs of broken ends fuse with each other. (See the drawing on p. 82.) In this manner, the dicentrism is perpetuated. In cells throughout the organism, this cycle of breakage, fusion, and bridge formation is repeated a number of times within the lifetime of the plant, ending when the broken ends eventually heal without fusing; but in the endosperm tissue (the kernel), it appears to repeat indefinitely. The breakage-fusion-bridge cycle results in massive mutation, revealed in characteristic patterns of variegation of the resulting endosperm tissue. Many of these mutations had never been seen before, and some of them involved gross changes in the arrangement of the chromosome.

This work was reported in a series of papers, beginning in 1938. It was of interest for several reasons. To some people, it contained the proof they needed that the rejoining of chromosomes was not a random event, but, rather, the result of highly specific forces governing chromosomal interactions. To others, it was of primary interest as a way to explain the origin of large-scale mutations. To McClintock it was both; it was also evidence of yet another mechanism the organism had evolved for generating change.

But success in research did not translate into institutional success. The position at Missouri, seemingly a solution to so many difficulties, did not work out. The question is: Why not? Why is it that after five years she was once again at loose ends?

McClintock's own view is that her days at Missouri were numbered from the start. Early on, she concluded that the job had been "specially created" for her by Stadler. "It was very special and very good—I had privileges that most others didn't have *but* I realized after I'd been there a couple of months that this was no place I could ever really stay. I had that disability that I was always alone. I had no chance of being promoted, I was excluded from faculty meetings and things like that, I had no real part." In the course of time, the discomforts of her professional position generated increasing distress. While her reputation in the world of genetics continued to grow (in 1939 she was elected Vice President of the Genetics Society of America), her status at the University of Missouri did not. What she found more irritating yet was the failure of her department to notify her of job offers from other institutions. "Letters would come along, inquiring about people ('Did they have such and such a

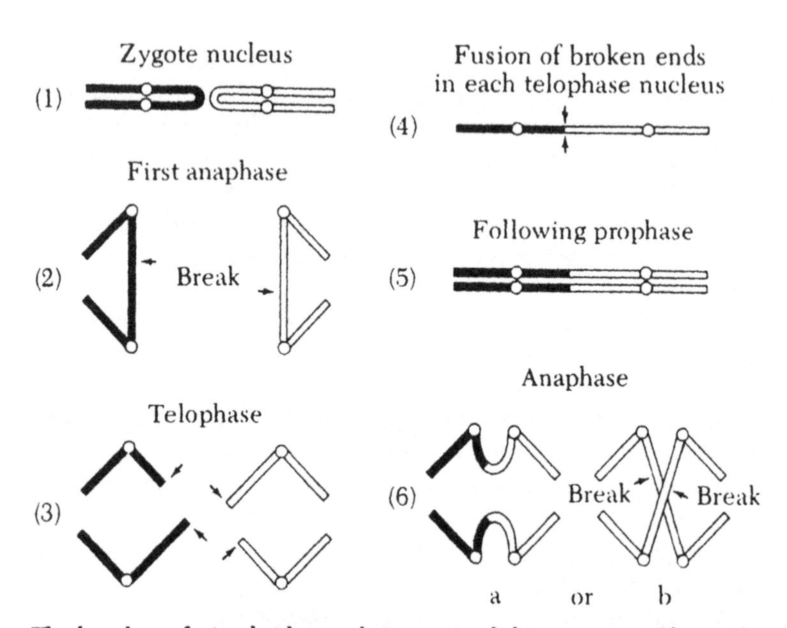

The breakage-fusion-bridge cycle in a pair of chromosomes. This cycle may be initiated if each gamete contributes a chromosome that has been broken in the anaphase of the division preceding gamete formation. The zygote nucleus will then contain two such chromosomes. In the prophase of the first division of the zygote, each of these is composed of two sister chromatids fused at the position of the previous anaphase break (1). In the first anaphase of the zygotic division, these two chromosomes give rise to bridge configurations as the centromeres of the sister chromatids pass to opposite poles (2). Breaks occur in each bridge at some position between the centromeres. In the telophase nuclei, two chromosomes, each with a newly broken end, are present as diagrammed in (3). The crosses mark the broken ends of each chromosome. Fusion of broken ends of chromosomes occurs in each telophase nucleus, establishing a dicentric chromosome (4). In the next prophase (5), several types of configurations may result from separation of the sister centromeres, two of which are shown in (6). Separations as shown in (6b) give rise to anaphase bridge configurations. Breaks occur in each bridge at some position between the centromeres. The subsequent behavior of the broken ends, from telophase to telophase, is the same as that given in diagrams (1) through (6). After Barbara McClintock, "Chromosome Organization and Genic Expression," *Cold Spring Harbor Symposia on Quantitative Biology* 16 (1951): 14.

person?') and they were not directed to me. Then, later, letters would come asking why I wasn't interested, and saying they were told I was to be promoted. But I was never even told anything about it."

Nor was she promoted. But they did, she thought, want her to stay. "I just knew it was impossible. They were always talking about the people they were hiring as associate professors and so forth, people whose credentials weren't up to what I had. And I was still an assistant professor. They would talk to me about the hiring, assuming it didn't matter to me. Finally, the time came when I thought, 'This is it!' So I went into the Dean's office and asked, 'What are my chances here?' He told me that if anything happened to Stadler, they would probably have to fire me." Whereupon she requested a leave of absence, without pay, and left Missouri. This was June 1941.

Just why she was given so little encouragement to stay is difficult to say. Almost certainly, an explanation based on simple discrimination on account of her gender does not suffice. Missouri had not only hired her in the first place, but had in fact promoted another woman (Mary Jane Guthrie) to the rank of associate professor of zoology during McClintock's tenure. Was it, then, that she was perceived as "difficult" because she demanded—and resented the absence of—rights commensurate with her merit? To some degree, probably yes. Five years as an assistant professor had not only failed to socialize her, but had, if anything, made her seem even more problematic. She had long ago rejected the restricted role of a "lady scientist"; but when she had a title normally reserved for a gentleman academic, she did not fit that role either. Just how much her own failure to accommodate was provoked by Missouri's failure to accord her the same administrative consideration as her colleagues must remain an open question, but the fact remains that by now the rules and conventions of her environment had less meaning to her than ever before.

As she came to realize later, she would "do things others didn't do—I never thought anything about it." One such incident came to be known all over campus (many years later she was shown photographs that someone had taken of it): "One Sunday, I arrived without my keys and so I climbed up the side of the building and let myself in the window." To her, it seemed the most obvious thing to do.

She recalls other, less amusing incidents that also irked the authorities. There was a rule stating that graduate students were not allowed to stay in the laboratory after 11:00 at night. But she thought her graduate assistant's work was important, so she said, "Go ahead and work." Or, if she felt that a particular student would do better elsewhere, she would encourage that student to leave Missouri. Institutional loyalty was clearly not a high priority with her.

She tells about another problem that she also thinks might have gotten her into trouble. In the first years of her appointment, Cornell continued to serve as a kind of home base. That is where she would return every summer to grow her corn. After gathering the mature corn, at the end of the summer, she would come back to Missouri, usually in time for the start of classes. Sometimes it happened, however, that the corn would be late in maturing, and "I would just have to wait. There was no need actually for me to get back, I would have everything arranged." A formal calendar was just one of the many rules she regarded as unimportant. Indeed, she tells these stories with unmistakable pride, as illustrations of her independence from foolish conventions. To others, however, such behavior must have seemed provocative. The authorities did regard the rules as important, and they frequently called her down for such infractions.

Escapades, absentmindedness, or irreverence are the stuff out of which scientific folklore is made. It is a half-truism that idiosyncrasy may be the price paid for originality, and most university communities have become accustomed to tolerating a certain amount of eccentricity on the part of their most brilliant scientists. But in McClintock's case, the tolerance that was certainly felt by some of her colleagues seems finally to have been outweighed by more negative feelings.

She believes that a man would have gotten away with most of the things she did, but that being a woman *and* a maverick was simply too much. In fact, she thinks that intolerance of each spilled over onto the other: "A good excuse, one for the other." Undoubtedly, she is at least partly right. A man would probably have been penalized less harshly. But more to the point of the story, the fact of being a woman reinforced her eccentricities, both in others' perceptions and in fact. Her alternatives, as a woman, were simple: be a "lady" or a "maverick." In her own terms, she had no choice. Her decision to pursue what gave her most pleasure had become a commitment. And around it, she had had to make rules for her own existence—rules that had by now become a source of pride, and just as binding as those of the conventions she rejected. They were an essential part of her self-definition—visible evidence of her difference.

A related but somewhat separate concern was her forthright manner of speech. No doubt some of her colleagues felt stung by her sharp wit. Even at Cornell, the combination of her own exceedingly quick mind and her corresponding impatience with others less quick than she had led to some resentments. Now, ten years later, she was no less free-speaking and probably a good deal more conscious of her own abilities. Even the brilliant Lewis Stadler came in for some of her acerbic remarks. However much she respected him as a theoretician, as an experimentalist he fell far short of her own high standards, and she didn't hesitate to say so. Stadler himself seems not to have minded, but others did.

By the summer of 1940, the Dean of Liberal Arts, W. C. Curtis, had already decided that it was no longer in the interests of Missouri to keep McClintock on the faculty. The notes of Frank Blair Hanson of the Rockefeller Foundation tell of a conversation he had had with Curtis in Woods Hole that summer. Hanson reports Curtis as saying: " 'Barbara McClintock has proved to be a trouble-maker and Curtis hopes an offer may come her way so that she can have her career elsewhere than at Missouri.' "[3]

Just a year later, two months after McClintock's departure on leave of absence, the Dean got wind of a rumor that she had been nominated for membership in the National Academy of Sciences. "Trouble-maker" or not, she remembers his trying to woo her back, promising a raise and a promotion.[4] "But by that time I had gone." Once her decision was made, it would not be reconsidered; she was disenchanted with university life. "It just meant that there was no hope for a maverick like me to ever be at a university."

And if there was no hope for her at a university, where then? In the mid-1930s, when she'd been jobless before, she'd begun to think about alternatives to a career in genetics. Now once again she looked around for other things to do. This move was not simply pragmatic; it was part of a lifelong strategy she had developed for dealing with depression. "I get into something else, some new knowledge. . . . It takes me away from brooding about myself." At the time it became clear she could no longer remain at Missouri, she remembers considering meteorology as a possible new career. But it was never a serious enough impulse to actually take her away from genetics.

The professional dislocation and the lack of institutional support she experienced throughout the 1930s clearly took a serious toll on her morale. But it did not undermine her basic commitment to the work that had captured her deepest interests. In part this can be attributed to the fact that that work had taken on a life of its own. In return for the emotional and intellectual energy she invested in it, such a life provided its own sources of gratification, compensating even for the frustrations of life lived less happily in other spheres. It offered, as Einstein once wrote, "the peace and serenity that [one] cannot find with the narrow confines of swirling, personal experience"[5] as well as a form of intimate communication with a piece of nature far from the realm of human exchange.

In addition, it generated a form of personal affirmation. In spite of all the institutional difficulties she experienced and the interpersonal conflicts these reflected, she nevertheless retained the enthusiastic support of her most respected colleagues. However anomalous her position was professionally, she suffered neither intellectual isolation nor rejection during this period. If anything, her institutional marginality served to strengthen her intellectual and emotional commitment to her work.

The absence of a professional niche did, however, have longer-term consequences. As a mature scientist, McClintock had a style of research all her own. The questions she asked, and the explanations or "understanding" she sought, were not quite the same as those of her colleagues. Perhaps she would have resisted the forces of professional socialization that normally make for intellectual conformity no matter what her position, but almost certainly, the anomalies of the position she held served to exacerbate the differences that existed. Not sharing in either the rewards or responsibilities that go with a conventional scientific career, she was freer than most to cultivate her own interests and inclinations. Her style of research became more and more her own, less and less subject to influence by current trends. In the 1930s, the differences between her intellectual commitments and those of her colleagues were still readily bridged. There were no serious barriers to communication of the kind that were to emerge later on.

It would nonetheless be useful at this point to examine those differences, partly in order better to understand the nature of the breach that eventually did occur, and partly to gain some insight into the concerns that guided her work in general.

Barbara McClintock was then, as she is now, primarily a cytologist, but she was also, at the same time, a geneticist, a naturalist, and more. To clarify how her interests stood in relation to those of her colleagues and how such eclecticism affected her work, the next chapter briefly reviews the relations of genetics and cytology, not only to each other, but also to the other principal biological disciplines of the time.

CHAPTER 6

Interlude: A Sketch of the Terrain

In *A Short History of Genetics*, L. C. Dunn has described the ten years preceding the outbreak of World War II as the "climactic" decade of classical genetics. By the early 1930s the chromosomal basis of genetics was fully established, but the mechanics were still unknown. The fundamental questions of the time were still: What is the nature of the gene? Of genetic transmission? Of variation? Of mutation? In the 1930s cytogenetics provided the principal means of addressing these questions. A great boost to cytological analysis came from the discovery of the giant salivary gland chromosomes in *Drosophila* (1933). Chromosomes could be clearly observed and tracked in the two best studied organisms of genetics—*Drosophila* and maize. By focusing attention on the physical basis of genetic events, the study of chromosomes and of the internal mechanics of chromosome variation prepared the way for the transition from the formal genetics of the classical period to the molecular genetics of the modern period.

But the gene was not the only, or even the principal, object of interest to cytologists. Genes, after all, were not what one saw; one saw chromosomes and their parts. The gene was at that time, as in many ways it still is, an abstract construct; to the cytologist, the chromosomes were the "real things"—of interest in and of themselves. In the opening pages of his classic text published in 1937 (*Recent Advances in Cytology*), the British cytologist C. D. Darlington reminded his colleagues that "while the chromosomes contain the 'something' which we identify with the genotype, they themselves cannot be directly identified with it."[1] Even twenty-seven years later, when DNA had long since been identified as the genetic material and therefore as the crucial component of chromosomes, Darlington illustrates a perspective likely to be overlooked by molecular geneticists. In the opening address to the First Oxford Chromosome Conference (1964), he explains:

> For me it means that the sequence, molecule-gene-chromosome-organism-community, is a physical hierarchy corresponding to a living hierarchy, an adaptive hierarchy. In evolving this adaptive hierarchy—as in the sexual cycle—chromosomes always come first; organisms come second.
>
> How different from this view of ours are the views of our colleagues in kindred sciences! They have heard of chromosomes, but to describe them they use a different language. The *anatomist* thinks of the chromosomes as very small rod-shaped bodies possibly present in all cells, visible with difficulty during the process of cell division and believed to contain the hereditary factors.
>
> The *chemist* is more definite, and sees the chromosomes as representing a chemical structure and a genetic code which he has himself discovered. The structure and the code together place the chromosome in his total scheme of nature: from them the chromosome, as well as all the rest, can be deduced. The *experimental breeder* approaches the question from the other side. He can tell us what the chromosomes are like from looking at the whole organism. For him the chromosome contains a group of genes identified by, and deduced from, his own experiments, and enabling him to predict what will happen in whole

plants and animals. The *mathematical geneticist* formalises these notions; he sees the chromosome and its constituent genes as a mechanical model obeying rules of recombination and mutation, interaction and selection. These rules he knows how to express in mathematical terms and from them he can ascertain the laws of evolution, laws derived from, and applied to, whole organisms. Lastly, the *naturalist* brings us back to the view of the anatomist; he sees the chromosomes as properties belonging to whole organisms which he knows. . . .

Thus all our colleagues, who are busy either with the structure of molecules or with the appearance of organisms, regard the chromosome as doing what various theories, such as the chemical theory of the chromosome and the chromosome theory of heredity, require it to do. They find that it is doing its job, or seems to be doing its job, smoothly and well—so smoothly and so well that they can take it for granted; they can deduce its properties; they do not need to observe them.

We must applaud the success achieved by our colleagues on the basis of these assumptions. But they see the chromosomes through the mind's eye. We, who believe we see actual chromosomes through the microscope, must explain what we have seen, and point out that it is not always what our friends expect.

For us, neither the chemical code, nor the linkage map of the chromosome, nor the genes embodied in it, are enough.[2]

Darlington's review is especially pertinent to an understanding of Barbara McClintock's work and thought as it evolved through the 1930s and 1940s. Genetics in the 1930s did not yet hold sway over other biological disciplines to the extent that it has in more recent times. The relatively independent perspectives of the cytologist, of the developmental embryologist, or the naturalist, all represented strong competing interests. Evolutionary concerns were only beginning to be integrated with those of genetics.

It is no easy matter to place McClintock on Darlington's or, for that matter, anyone's map. In a way that has always given her a special advantage, her manner of thinking has been marked by multiple commitments, or, more accurately, by a relative lack of commitment to any one school. We have already seen how productive her ability to look at maize from the dual perspectives of the geneticist and the cytologist was for her work in the late 1920s. Yet a third strand comes into view in her work from the early 1930s on the Nucleolar Organizer Region. Here, her emphasis on the concept of organization, and even her use of the term "organizer," suggests that some influence from embryology was by then also at work.[3] In the years to come, her awareness of the interests and concepts of the embryologist (or developmental biologist) would prove to be an even more fertile source of ideas. And finally, during the years she spent at Missouri, one can begin to see certain features in her relation to her work which, although probably personal in origin, are strongly reminiscent of the naturalist tradition.

In the 1930s, her combination of interests was an unusual one; in the 1940s and 1950s, it became even more so. The growing success of Mendelian genetics, by then grounded in a vast body of cytogenetic evidence, was changing the way in which biologists thought. It shifted attention away from the organism as a whole, concentrating interest more and more on the properties and behavior of individual genes. To the extent that other disciplines (such as embryology) did not share in this focus, their work came to seem more and more remote from genetics. And as the predictive and explanatory success of genetics accumulated, its concerns, interests, and methodology came to serve as a model of what a biological science should be. How the rise of genetics affected the alignment of loyalties, goals, and interests in biology as a whole is a story not yet fully understood, but some of its principal features can be identified.

Broadly speaking, the three major concerns of twentieth-century biology have been, and still are, heredity, development, and evolution. But the relation between these three concerns has changed dramatically over the course of the century. In the latter half of the nineteenth and early twentieth centuries, heredity and development were one science. Any theory of heredity that did not include an accounting of developmental phenomena—a theory, for example, that focused solely on the mechanism of genetic transmission—would have seemed entirely inadequate to students of inheritance and development alike. But with the birth of the chromosomal theory, genetics and development began to diverge. As the concerns of geneticists focused more and more on the nature of the gene and the mechanism of its transmission from generation to generation, those of the developmental biologist came to focus on questions of organization and on the determinants of growth from egg to mature organism. Indeed, it was far from obvious to the developmentalist that genes were at all involved in these events. Equally, if not more, important might be factors in the cytoplasm. The precise nature of the gene, and the details of how it worked, accordingly seemed to be "incidental and secondary" concerns.[4]

Furthermore, the new science of genetics exposed a fundamental paradox dividing genetics from developmental embryology—a fact to which Morgan himself called attention. In 1910, on the eve of his conversion to Mendelian theory, he wrote:

> If Mendelian characters are due to the presence or absence of a specific chromosome, *as Sutton's hypothesis assumes* [italics in original], how can we account for the fact that the tissue and organs of an animal differ from each other when all contain the same chromosome complex?[5]

At the time Morgan, who was originally an embryologist, saw in this a grievous objection to the chromosome theory; once he was converted, the problems of development faded from his attention. (Twenty-four years later, he returned to the problem and attempted to bring the two fields together in his 1934 book, *Embryology and Genetics*.[6] But the hope of synthesis was premature. When confronted by one disappointed reader, he is reported to have replied, "I did just what the title of the book suggests. I discussed embryology *and* I discussed genetics."[7])

In the mid-1930s, genetics and development were nowhere near a rapprochement. Practitioners in each field were largely ignorant of the details of the other and uninterested in, if not scornful of, one another's preoccupations. Fundamental methodological and philosophical commitments divided them yet further. The theory of the gene was quintessentially mechanistic and the methodology of genetics inherently quantitative; geneticists were entranced with the power of numbers. By contrast, embryology was a more qualitative science, concerned with overall shape and form, and inevitably more preoccupied with the individual organism than genetics needed to be.

The need to relate the two fields was nevertheless evident to all, and some embryologists worried about the effect the geneticists would have on their field. Ross Harrison of Yale University, retiring from the vice presidency of the American Association for the Advancement of Science in 1937, sounded a warning:

> The development of gene theory is one of the most spectacular and amazing achievements of biology in our times. The embryologist however is concerned more with the larger changes in the whole organism ... than with the lesser qualities known to be associated with genic action.
>
> Now that the necessity of relating the data of genetics to embryology is generally recognized and the "wanderlust" of geneticists is beginning to urge them in our direction, it may not be inappropriate to point out a danger in this threatened invasion.
>
> The prestige of success enjoyed by the gene theory might easily become a hindrance to the understanding of development by directing our attention solely to the genome, whereas cell movements, differentiation and, in fact, all developmental processes are actually effected by the cytoplasm.... Such theories are altogether too one-sided.[8]

• • •

If embryology and genetics seemed to be at cross purposes, so did genetics and evolution. Indeed, the early relations between the latter two show certain parallels with the relations between the former. To the first generation of Mendelians, the theory of natural selection seemed entirely inadequate as a means of accounting for evolutionary change. Not until the 1930s did a successful integration of genetics and evolution become possible. The crucial point was that evolution is something that happens to populations; the proper focus of the study of evolutionary change is therefore the *distribution* of genetic traits in a population. Out of this recognition, Haldane, Fisher, and Wright (among others) developed a mathematical theory of population genetics that became the basis of a grand synthesis between such previously diverse fields as genetics, biometrics, paleontology, and systematics.

Initially, controversy over Darwinian theory among geneticists centered on two basic issues: one was the question of the directedness of small-scale evolutionary change, and the other, the question of the role of selection in the emergence of new species. The first question was settled relatively early in the minds of most workers in the field. Studies of the phenomenon of mutation seemed to make a belief in the inheritance of environmentally directed changes (acquired characters) superfluous. It was simpler to assume, as the neo-Darwinians did, that the gene changes that give rise to evolution are random. The apparent directedness of evolutionary change results from the operation of natural selection on variations produced by spontaneous mutations. Pockets of resistance to the radical neo-Darwinian stance, and even of belief in the direct inheritance of acquired characters, persisted for some time to come among a few geneticists and, more pervasively, in those areas that were conceptually remote from Mendelian genetics.

The question of the role of selection in the emergence of new species was more generally troublesome. Until the 1930s, quite a large number of geneticists remained resistant to the Darwinian notion that natural selection, acting on small changes in individual organisms, could be the driving force of evolution. Large (or macro) mutations were thought to be necessary to account for the origin of new species. But almost all remaining doubts seemed to be laid to rest by the development of population genetics. New species, in this view, result from the slow and gradual accumulation of mutations within an ancestral population that had been isolated by a geographic barrier. Even though this claim is now being challenged in some quarters, it provided a satisfying answer to a question that had long plagued biologists and allowed for a final acceptance of Darwinian theory by most geneticists. As Ernst Mayr recently wrote:

> The new synthesis is characterized by the complete rejection of the inheritance of acquired characters, an emphasis on the gradualness of evolution as a central tenet of Darwinian theory, the realization that evolutionary phenomena are population phenomena, and a reaffirmation of the overwhelming importance of natural selection.[9]

With genetics and evolution apparently reconciled, biologists could feel a far deeper confidence in their theories about both.

• • •

The immediate effect of this synthesis was that the domain and influence of genetics were vastly extended. But in science, as elsewhere, success and orthodoxy have a natural kinship. As the success of genetics continued to grow, so did the delineation of "right doctrine." The very real and substantial advances of classical genetics promoted confidence in an entire set of methodological and philosophical postures congenial to the new science—postures that affected conceptions of what questions were important, of what constituted adequate answers to such questions, and of how best to obtain such answers. In turn, the entrenchment of these particular methodological and philosophical commitments led to particular kinds of scientific advance.

The task of analyzing this mutual interaction is surely one of the most difficult facing contemporary historians of science. One of the many questions that needs to be asked is: What inclines an individual scientist to a particular set of such methodological and philosophical commitments, to resisting or accepting the dominant trend within a field? The answer must certainly depend on a wide range of factors, extending from the purely scientific to the social and psychological. And if Barbara McClintock maintained a methodological and philosophical independence, resisting the commitments that grew increasingly dominant within her own field, surely that independence had some relation to her particular "idiosyncracies of autobiography and personality."[10]

Her views of what was important were often at variance with those around her. She was duly skeptical of those who "thought they were going to solve the genome." To her, the gene wasn't solvable—it was "merely a symbol." "We were using a set of symbols in just the same way that a physicist used symbols." Accordingly, she was even more dubious about the new synthesis between genetics and evolution—population genetics. The entire analysis, she felt, was based on "inadequate concepts." Population genetics "were dealing with entities that were symbols, and these symbols were not good enough to handle in the way they were handled." More generally, she was critical of the zeal geneticists had for quantitative analysis. They were "so intent on making everything numerical" that they frequently missed seeing what was there to be seen. Her own method was to "see one kernel [of corn] that was different, and make that understandable." She felt that her colleagues, in their enthusiasm for "counting," too often overlooked that single aberrant kernel. And in line with all these other differences was her early and ongoing interest in embryology. To most geneticists, the work of cytogenetics led toward an increasing preoccupation with the physical-chemical dimensions of the problem of genetic transmission and variation; this emphasis simultaneously prepared the road to molecular genetics and diverted attention from the problems of development. This was not true for McClintock. Despite her preoccupation with the precise mechanical details of cytogenetic processes, because of her commitment to the whole organism, she never lost her interest in developmental problems. That interest may not have found direct expression in her published work of the 1930s (except perhaps in her work on the Nucleolar Organizer Region), but it played critically into the radical formulation she developed in the 1940s.

Even within genetics, McClintock was not entirely alone in her concerns. Other geneticists were also interested in embryology, even if they didn't employ it in their actual work. A number of them were vocal critics of the reductionist trend in modern genetics. McClintock was far from being the severest critic in her field. Almost certainly, that distinction belongs to Richard B. Goldschmidt, the most notorious dissenting geneticist of his time. Because of his extreme unorthodoxy, Goldschmidt provides a useful vantage point from which to appreciate the range of orthodoxy of the time. He also serves as a foil to McClintock. The two shared a number of common interests, including a deep respect for embryology and a skepticism toward prevailing dogma. But important differences in style divided them.

Goldschmidt came to America in 1936, the same year that McClintock went to Missouri. Hitler was in power, and he finally felt it was necessary to leave his post at the Kaiser Wilhelm Institute in Berlin. A new home was propitiously offered him at the University of California in Berkeley. For many years Goldschmidt had been a relentless critic of chromosomal-Mendelian genetics. His work in developmental physiology and natural history convinced him that the concept of the gene as a unitary element, a "bead on a string," was scientifically as well as philosophically inadequate. By the mid-1930s, a crucially troublesome new phenomenon had been firmly established: the "position effect" in *Drosophila,* a term coined by Alfred Sturtevant. The phenotype expression of a particular gene (called "bar-eye" for the bar-shaped eye it gave rise to) had been shown to depend on its relative position on the chromosome. For Goldschmidt, the "position effect" constituted final proof that a new interpretation of genetics—radically different from that of the Morgan school—was absolutely essential. He arrived in America proclaiming: "The theory of the gene is—dead!"[11]

In place of the static classical theory, Goldschmidt offered a global and more dynamic theory, which dispensed with the notion of individual genes as separate units. Instead, he proposed to take the chromosome as a whole as the agent of genetic control. Genetic changes might be related to (more or less) specific *sites* on the chromosome, but they result, he argued, from a rearrangement of parts that affect the functioning of the chromosome as a whole. In his formulation, "macromutations"—chromosomal rearrangements that have large-scale consequences for the organism—could be readily visualized, and the conceptual difficulty in explaining the origin of new species overcome. In the same scheme, a scheme that might be regarded as a highly speculative precursor to McClintock's later interpretation of genetic organization, Goldschmidt saw a solution to the problems of development as well. Development, he reasoned, could be regulated by structural changes that lead to the activation (or expression) of different segments of the chromosomes at different times.

Goldschmidt's provocative challenges were not well received by his American colleagues, some of whom regarded him as a mere "obstructionist."[12] In part, he laid himself open to censure, and ultimately to total rejection, by a paucity of supporting observations for his arguments. At best these arguments were seen as "philosophical," at worst as simply "wrong." His genetic theory directed attention to some major difficulties with the classical gene concept, but it had too weak a basis in experimental fact to survive. In time, the concepts of point mutations and individual genes were amply confirmed by observation, and on a number of basic questions, Goldschmidt's position became simply untenable. Today, on the other hand, evidence for chromosomal rearrangements is mounting, and the argument for the necessity of discontinuous variation to account for the origin of new species is being resurrected. In hindsight it appears that the latter argument, at least, was too hastily rejected. Stephen Jay Gould wrote: ". . . the [neo-Darwinian] synthetic theory made a caricature of Goldschmidt in establishing him as their 'whipping boy.' "[13] To the end of his life Goldschmidt remained an unrepentant critic of genetic theory, and when he died in 1958, after a long and illustrious career as an Old World biologist, he was an outcast from the community of modern geneticists.

Barbara McClintock offers us a much subtler version of heterodoxy. Unlike Goldschmidt, McClintock's work placed her in the mainstream of cytogenetic research. She belonged fully to the age of Mendelian genetics and had herself contributed significantly to establishing the chromosomal basis of Mendelian theory. Nevertheless, she shared a number of Goldschmidt's interests and reservations. She admired his critical capacity and maintained a similar skepticism toward some of the assumptions of her colleagues, most notably on evolutionary questions.

Temperamentally, the two could hardly have been more different. Where Goldschmidt was flamboyant, she was reserved; where he was extravagant, she was cautious. Many of Goldschmidt's theories were based on inadequate evidence; by contrast, McClintock's commitment to the demands of experimental proof was unshakable. Her papers were characterized by a cautiousness of interpretation and meticulous regard for observational proof that upheld the highest standards of the new biology. (In fact, Marcus Rhoades remarked that he used her papers as teaching models of scientific clarity and rigor.) For these reasons, McClintock was not vulnerable to the same kinds of criticism that Goldschmidt received. She was highly critical herself of his proposals for a new genetic theory—"they were made out of ignorance"—but she thought he was right about evolution. And she respected his courage and his ability to identify problems others did not see or were willing to ignore.

Despite her own critical and iconoclastic leanings, McClintock shared enough of the values of her community during this period to avoid the kind of conflict that Goldschmidt ran into. For the most part, she kept her criticisms to herself, sharing them, if at all, only with close friends. Her unconventionality expressed itself more in the style and focus of her work than it did in direct theoretical controversy.

In her mid-thirties, Barbara McClintock's particular scientific style was already well defined and emerging in ever-sharper relief. Its distinctive features were polar in character: its ultimate strength derived from a dialectic between two opposing tendencies. The reader may recall from the previous chapter a characterization of her work by Morgan as "highly specialized." And, although few would accept Morgan's description of the cytology of maize genetics as any more narrow a category than the cytology of *Drosophila* genetics, one aspect of McClintock's scientific preoccupations may easily have led him to such a description: her focus on the minutest of details. The tenacity with which she hunted down every observable chromosomal modification, the thoroughness and rigor that accompanied her virtuoso technique—all these might lead one to think of the focus of her search as narrow. In fact, what she consistently pursued was nothing less than an understanding of the entire organism.

The word "understanding," and the particular meaning she attributed to it, is the cornerstone of Barbara McClintock's entire approach to science. For her, the smallest details provided the keys to the larger whole. It was her conviction that the closer her focus, the greater her attention to individual detail, to the unique characteristics of a single plant, of a single kernel, of a single chromosome, the more she could learn about the general principles by which the maize plant as a whole was organized, the better her "feeling for the organism."

To some extent the stories she tells to illustrate the meaning of "understanding" recall what might be called the "naturalist" tradition—a tradition that in most parts of biology had long been replaced by the experimental tradition, but traces of which nevertheless still survived in the mid-1930s. McClintock was able to incorporate and integrate these traces into an otherwise thoroughly experimental stance. No scientist ever develops in a vacuum, but it is difficult to find any direct intellectual influences that can be held responsible for this element in her thought. Rather, it would seem that she came to this amalgam in her own highly individualistic way, dictated more by internal forces than by external ones. The stories themselves reflect this.

She describes, for example, an experience from her early days at Missouri when she was pursuing some elaborations of the ring chromosome motif. "Most of the time, ring chromosomes undergo semiconservative replication, but there are occasional sister strand exchanges which will produce a double-size ring, having two centromeres," she says. "At anaphase, [the two centromeres] start to go to opposite poles, but [since] it's a double-size ring, it breaks. It breaks in different places, and the broken ends fuse to form new rings—different from the old. If they are small, they will frequently get lost—they will not make the telophase. Among the plants that were growing in the culture I was dealing with, there were those that could have one, two, or three of these rings, in any combination. One ring might be small; the other two could be a bit larger. In these plants, the ring chromosomes carried the dominant genes, and in order to get the recessive expression, the ring carrying the gene would have to get lost."

So adept did she become at recognizing the outward signs of those structural alterations in chromosomal composition that she could simply look at the plants themselves and know what the microscopic inspection of the cells' nuclei would later reveal. "Before examining the chromosomes, I went through the field and made my guess for every plant as to what kind of rings it would have—would it have one, two, or three, small or large, which combination? And *I never made a mistake,* except once. When I examined that one plant I was in agony. I raced right down to the field. It was wrong; it didn't say what the notebook said it should be! I found that . . . I had written the number from the plant adjacent, which I had not cut open. And then everything was all right."

Her belief is that the mind functions "like a computer"—processing and integrating data far more complex than we can possibly be conscious of. And finding the cause of her mistake —a misfiling error, as it were—was reassuring. "That made me feel perfect, because it showed me that whatever this computer was doing, it was doing it right." All she was conscious of doing was "looking at these fine stripes of recessive tissue"; she says the computer did the rest. "And I never made a mistake." The crucial point of this story, to her, is the state of mind required in making such judgments. "It is done with complete confidence, complete understanding. I understood every plant. Without being able to know what it was I was integrating, I *understood* the phenotype." What does understanding mean here? "It means that I was using a computer that was working very rapidly and very perfectly. I couldn't train anyone to do that."

Since her days as a young graduate student, she had always carried out the most laborious parts of her investigations herself, leaving none of the labor, however onerous or routine, to others. In this she did as almost all beginning scientists do. But most scientists, as they mature, learn to delegate more and more of the routine work to others. There are, of course, exceptions, and Emerson, who, according to Harriet Creighton, "didn't regard anything as routine," was one. He prided himself on doing his own work. For McClintock, more than pride was involved. Her virtuosity resided in her capacity to observe, and to process and interpret what she observed. As she grew older, it became less and less possible to delegate any part of her work; she was developing skills that she could hardly identify herself, much less impart to others.

The nature of insight in science, as elsewhere, is notoriously elusive. And almost all great scientists—those who learn to cultivate insight—learn also to respect its mysterious workings. It is here that their rationality finds its own limits. In defying rational explanation, the process of creative insight inspires awe in those who experience it. They come to know, trust, and value it.

"When you suddenly see the problem, something happens that you have the answer—before you are able to put it into words. It is all done subconsciously. This has happened too many times to me, and I know when to take it seriously. I'm so absolutely sure. I don't talk about it, I don't have to tell anybody about it, I'm just *sure* this is it."

This confidence was not new. She tells of an occasion from her early days at Cornell—another instance in which "understanding" seemed to bypass any conscious awareness: "A particular plant was heterozygous with respect to a translocation; that is, one chromosome carried the translocation, while the homologous chromosome was normal. According to meiotic segregation, it would normally give pollen grains which would be 50 percent defective [hence sterile] and 50 percent normal. We had at that time a post doc who had just started working with translocations. He was examining pollen from these plants, expecting them to be either perfect [having no translocations] or to be heterozygous [where 50 percent of the pollen would be sterile]. He came to me and said, 'There are some plants that are 25 to 30 percent sterile, *not* 50 percent sterile.' He told me this in the field and he was disturbed." McClintock was disturbed too—so much so that she left the field, down in the hollow, and walked up to her laboratory. There she sat for about thirty minutes, "just thinking about it, and suddenly I jumped up and ran down to the field. At the top of the field (everyone else was down at the bottom) I shouted, 'Eureka, I have it! I have the answer! I know what this 30 percent sterility is.' " When she got to the bottom of the hollow, the group of corn geneticists working there gathered around her, and she realized she couldn't provide the reasoning behind her insight. "Prove it," they said. "I sat down with a paper bag and a pencil and I started from scratch, which I had not done at all in my laboratory. It had all been done fast; the answer came, and I'd run. Now I worked it out step by step—it was an intricate series of steps—and I came out with what it was. [The post doc] looked at the material and it was exactly as I'd said it was, and it worked exactly as I'd diagrammed it. Now, why did I know, without having done a thing on paper? Why was I so sure that I could tell them with such excitement and just say, 'Eureka, I solved it!'?"

Perhaps the answer, once again, depends on the intimate and total knowledge she sought about each and every plant. A colleague once remarked that she could write the "autobiography" of each plant she worked with. Her respect for the unfathomable workings of the mind was matched by her regard for the complex workings of the plant, but she was confident that, with due attentiveness, she could trust the intuitions the one produced of the other. In the years to come, that confidence would become a source of vital sustenance.

CHAPTER 7

Cold Spring Harbor

I am a little piece of nature.

ALBERT EINSTEIN

Even after the United States entered World War II, civilian life on this side of the Atlantic continued to function much as usual. It would be many months before the war was felt in the lives of most Americans, and, unlike their colleagues in physics, biologists were to retain the freedom of relatively nonessential civilians throughout the war. World War II was not a biologists' war.

On the eve of Pearl Harbor, Barbara McClintock was preoccupied with a private upheaval. Once again, she was unemployed. In leaving Missouri, she left the only job she had, with no expectations of another. "I knew I could do something; I wouldn't starve." But beyond bread and butter, if she was to continue as a biologist, she needed a place where she could do her work, and a field in which to grow her corn. By this time, Cornell had long since ceased to be an option. Emerson was retiring, his students were widely scattered, and Esther Parker had bought a farm fifty miles north of Ithaca. All sense of Cornell as "a home to go to" had by now evaporated.

At such a time, one needs good friends, and Barbara McClintock had an especially dear and loyal one in Marcus Rhoades. She sent off a letter to him, when he had just taken a new position at Columbia University, to ask where he was going to grow his corn (presumably, it was not to be in Manhattan). An answer came promptly. Rhoades had not yet fully established himself, but that summer he was planning to go out to Cold Spring Harbor. "I'll go too," she thought, "and plant my corn there."

An invitation was easily arranged. She wrote to Milislav Demerec, a *Drosophila* geneticist she had known for many years. Demerec had been at Cold Spring Harbor since 1923, and, like most of his colleagues, he held McClintock's work in high esteem. She was a welcome guest. Forty years later she recalled, "I came in June, and when the summer was over, I stayed—I liked it very much here." But still, she had no job. She lived in one of the summer houses and managed to stay on until November, but, with winter coming on, the summer places finally had to be closed up. Fortunately, Marcus Rhoades had an extra room in his apartment at Columbia, and she went to stay there.

Not long after, on the first of December, 1941, Demerec became Director of the Department of Genetics of the Carnegie Institution of Washington at Cold Spring Harbor. One of his first moves in his new position of authority was to call McClintock and offer her a one-year position. She hesitated. She wasn't sure just what she wanted to do, and in her uncertainty, she felt disinclined to accept any formal obligation. But reality pressed hard. Persuaded finally by her friends at Columbia University, she accepted the job and returned to Cold Spring Harbor. After a couple of months, Demerec proposed making the appointment permanent.

"But even then," she remembers, "I hadn't made up my mind that I really wanted a job. I didn't know what I wanted to do. . . . I didn't want to commit myself to anything, because I enjoyed the freedom, and I didn't want to lose [it]." Demerec was insistent, and he urged her to go to Washington to speak with Vannevar Bush, then President of the Carnegie Institution —an obligatory procedure, she says, for a woman candidate: "Just go on down anyway. You can take the plane down and come back the same day," he said. Reluctantly, she agreed, "not caring whether I had a job or not." As it turned out, her meeting with Bush was decisive. In the spring of 1942, Bush, as Chairman of the National Advisory Committee for Aeronautics as well as President of the Carnegie Institution, was heavily involved in the war effort and had many other things on his mind. Nevertheless, he found time to see McClintock. "The fact is we had a marvelous time; we talked about a variety of things, and I was completely at ease. So was he." Before she got back to Cold Spring Harbor that evening, Bush had telephoned Demerec to support her appointment. "I took it, still not knowing whether I wanted a job. It was about four or five years before I really knew I was going to stay."

Why was it so hard for her to decide? The reasons are hard to fathom. The Carnegie Institution was offering her what no other institution had offered—a salary, a place to grow her corn, a laboratory for her research, and a home. Here was an environment in which she could pursue her own ideas and do her work as she saw fit, protected from departmental politics, from teaching duties, from administrative responsibilities. But perhaps too protected. In its idyllic seclusion, Cold Spring Harbor may have felt like an outpost: small, remote, and somewhat off the beaten track of scientific exchange.

The main year-round laboratories were sponsored by the Carnegie Institution of Washington—by then referred to as their Department of Genetics, but originally begun in 1904 with a handsome endowment as the Station for Experimental Evolution. Charles B. Davenport, the noted eugenist, was its first Director. With the help of half a dozen full-time resident investigators, Davenport had established the station as one of the main centers of early genetics research in America. After 1910 it came to be overshadowed by the laboratories at Columbia, Harvard, Cornell, and other academic research institutions, but it nevertheless continued to maintain its reputation as a significant center of new genetics research. The number of year-round investigators hovered between six and eight; these were joined by a small number of fellows and research assistants, and supported by a healthy number of assistants. In the summers, the influx of guests tripled the number of full-time investigators.

Occupying the same grounds was the Long Island Biological Association, an outgrowth of the summer laboratory of the Biology Department of the Brooklyn Institute of Arts and Sciences and supported (since 1924) by local residents. The Long Island Biological Association was primarily a summer research institute, attracting large numbers of distinguished geneticists from around the country.

On the North Shore of Long Island, Cold Spring Harbor combined the pleasures of a relaxed summer holiday with the stimulus of scientific exchange. With a small beach for playing, a "mess hall" for eating, and a few labs for working, it was an appealing place for biologists to spend their summers, with or without their families. In the summer of 1941, Cold Spring Harbor drew over sixty geneticists, many of their names well known from the development of classical genetics, and others, such as Max Delbrück and Salvador Luria, still young, whose fame would come with the building of a new genetics. Even the old Cornell group was represented by Barbara McClintock, Marcus Rhoades, and Harriet Creighton.

But with the arrival of autumn, the luminaries went home, leaving a small core of investigators to get on with their work. Even though it had a close connection with the Biology Department at Columbia, Cold Spring Harbor in winter was, by itself, too small to provide anything like the community of shared interests Barbara McClintock had known at Cornell. Even Missouri had offered greater possibilities for collegial exchange in her own field of work. With the exception of an occasional young researcher who would come for a brief period to work at Cold Spring Harbor (Peter A. Peterson was one), she was—

and would remain—the only maize geneticist there. There would be long winter evenings with no one to talk to, no like-minded soul to share ideas with, and perhaps above all, no one to joke with. Fondly remembering "Barbs's" earlier penchant for clowning, Harriet Creighton commented that McClintock just didn't have "enough fun" at Cold Spring Harbor. Although for others the community there might seem like one big family, it never seemed a family she belonged to.

Nevertheless, the benefits were undeniable. Despite her initial ambivalence, the Carnegie Institution provided a kind of patronage that may well have been her professional salvation. Wartime mobilization was about to cause a major upheaval in the nation's scientific establishment, and Cold Spring Harbor was, at the very worst, not too bad a place to be. Finally, there were no obvious available alternatives. Her work demanded a place, and early in 1942 she settled in at Cold Spring Harbor and resumed her investigations of the problems then of greatest interest to her—the chromosomal deficiencies that resulted from the repeated rounds of breakage and refusion. Once again the work she loved so much absorbed her energies and attention.

By the middle of that year the war was in full swing, and the Carnegie Institution was eager to articulate its own contribution to the general effort. The Board of Trustees resolved to "place war research first, and suspend its peacetime activities in order to do so," clearly recognizing the implications of this resolve. Vannevar Bush wrote, in the President's Report of June 1942: "To only a minor extent can we still hope to continue progress in paths of research towards distant cultural objectives." However, he continued, "Not all scientific talents are of such a nature as to be immediately and directly applicable to the waging of war, and hence the transition has occurred more rapidly in some departments than in others."[1]

Genetics was one of those departments with less immediate relevance than many other areas, and although some of the workers at Cold Spring Harbor shifted to wartime projects, for most, research continued as usual. Barbara McClintock's work was seen as being of importance to the "distant cultural objectives" of basic research, and as such it received much favorable attention in the annual administrative reports.

But even though research in genetics may not have been greatly affected by the war, daily life was. Signs of wear and tear were soon felt everywhere. The atmosphere at Cold Spring Harbor became even quieter than usual and distinctly more Spartan. The shortage of gas inhibited travel, rationing took a definite toll on the quality of food served in the dining room, and the influx of summer visitors dwindled to a trickle. There was little to do but work. As the months and then years wore on, new contributions in genetics poured out from Cold Spring Harbor and elsewhere.

Some of that work would eventually change the face of genetics, but few could then perceive the extent of the transformation that was in store. Research in what we now call the classical tradition continued vigorously and productively, with the investigations of such geneticists as McClintock adding more and more complexity to our understanding of inheritance. But the work of others was moving in a different direction—a direction that would soon lead to a picture of unimagined simplicity. In hindsight, most historians would say that the molecular revolution began in the early 1940s. The most important event by far was a discovery that took place just forty miles west of Cold Spring Harbor. In a laboratory overlooking the East River, Oswald T. Avery and his coworkers, Colin M. MacLeod and Maclyn McCarty—all of the Rockefeller Institute—found that DNA was the carrier of specific hereditary characters. But the full significance of that event could not yet be seen in the 1940s. The implications of other developments, subtler in their influence, were equally opaque. Who could then have predicted the ways in which work such as that of Luria and Delbrück (much of which actually took place at Cold Spring Harbor) would change the very questions geneticists asked? The 1940s were a time of ferment in biology, but the massive upheavals they set in motion would not be fully felt for another decade. And it would be many years before any of these developments had even an indirect bearing on McClintock's life and work.

In the meantime, McClintock's own investigations of anomalies in maize genetics were going very well. She produced a steady stream of new results that she described in the Annual Reports of the Carnegie Institution and in a major paper she wrote for *Genetics* ("The Relation of Homozygous Deficiencies to Mutations and Allelic Series in Maize"). Reference to her accomplishments figures prominently in the annual summary reports Demerec wrote for his department. Nevertheless, after two years at Cold Spring Harbor, a feeling of claustrophobia was mounting, and McClintock was ready for a break.

By good luck, in January 1944, a letter arrived from her old friend George Beadle suggesting that she come to Stanford for a visit. She responded immediately and enthusiastically. Cooped up with no cars, no money, and no friends, she felt a change was definitely called for. But the trip was not to be merely recreational; Beadle had a specific interest in bringing McClintock out to Stanford.

Three years earlier, he had impressed the biological world with his ingenious demonstration of a specific relation in *Neurospora* (a red mold on bread) between certain mutants and particular enzyme deficiencies, leading to his famous "one gene–one enzyme" hypothesis. But until now, analysis was limited by the fact that *Neurospora* was accessible only to genetic probing—its cytology had not yet been worked out. Indeed, *Neurospora* chromosomes were so small that they had eluded all previous attempts at identification. It seemed to Beadle that if there was anyone in the world who could crack this problem, it was McClintock.

Arrangements for the trip took a long time; not until late summer was she finally able to commit herself, assuming available space on the train, to a visit in mid-October.

That spring, America's most prestigious professional society —the National Academy of Sciences—elected Barbara McClintock as a member. It was the third time in their long history that a woman had been so honored. Florence Sabin had been the first (in 1925); Margaret Washburn the second (in 1931). Her friends were delighted; they felt that the honor was long overdue. Tracy Sonneborn, a geneticist with whom she shared a great deal of intellectual sympathy and mutual respect, was one of many who sent warm congratulations. She wrote back:

> It was both thoughtful and generous of you to write me as you did concerning the National Academy. I must admit I was stunned. Jews, women and Negroes are accustomed to discrimination and don't expect much. I am not a feminist, but I am always gratified when illogical barriers are broken—for Jews, women, Negroes, etc. It helps all of us.[2]

George Beadle—equally young for such an honor—was admitted in the same election. Just how these decisions were made necessarily remains obscure. The National Academy operates largely as a closed fraternity, with the existing members of a particular division responsible for the nomination and election of new members. Since records of these deliberations are not kept, nothing can be said either about the background of Barbara McClintock's election or about the rumored nomination of her three years earlier. The list of existing members is public, however, and a glance at that list—which included Stadler and Emerson—can leave no doubt as to who at least some of her supporters were.

Her election was cause for celebration: now she could go to Stanford as a visiting dignitary. Without question, it was a productive trip. Beadle later told Warren Weaver of the Rockefeller Foundation that "Barbara, in two months in Stanford, did more to clean up the cytology of *Neurospora* than all other cytological geneticists had done in all previous time on all forms of mold."[3] What appears in hindsight to have been a relatively straightforward piece of cytological analysis did not seem so at the time. A difficulty arose, and for us it is that difficulty which lends McClintock's experience in Beadle's lab its primary interest. Her own description of how she went about resolving it gives us some idea of how it was that she could "see so much more than others."

• • •

By her own account, her confidence had begun to fail even before setting out. "I was really quite petrified that maybe I was taking on more than I could really do." She went, set up the microscope, and proceeded to work, but after about three days, found she wasn't getting anywhere. "I got very discouraged, and realized that there was something wrong—something quite seriously wrong. I wasn't seeing things, I wasn't integrating, I wasn't getting things right at all. I was lost." Realizing she had to "do something" with herself, she set out for a walk.

A long winding driveway on the Stanford campus is framed by two rows of giant eucalyptus trees. Beneath these trees, she found a bench where she could sit and think. She sat for half an hour. "Suddenly I jumped up, I couldn't wait to get back to the laboratory. I knew I was going to solve it—everything was going to be all right."

She doesn't know quite what she did as she sat under those trees. She remembers she "let the tears roll a little," but mainly, "I must have done this very intense, subconscious thinking. And suddenly I knew everything was going to be just fine." It was. In five days, she had everything solved.

But what did she actually see when she went back to look through the microscope?

The problem to be solved was not simply to count and identify the chromosomes; the entire meiotic cycle of *Neurospora* lacked any description at this time. "There was great confusion about what kind of meiosis occurred in fungi, or even if there is anything typical." She did count the chromosomes; she found there are seven, distinguishable by their size and by their relative position. But her principal success lay in being able to pick out the chromosomes clearly enough to track them through the entire meiotic cycle. Her own description of what she had seen thirty-five years earlier still has a vivid narrative quality. It is easy to forget that she had to reconstruct the process from separate slides, that she did not see it unfolding in "live action." Even though the language is technical, it almost creates the illusion that we are watching it with her.

"The major thing I found out was that in the perithecium [fruiting body], where the asci [oval sacs containing the sexual zygotes] are going to be found, you get fusion of the nuclei from the two parents. Now what happens is that these nuclei go into prophase and then fuse following the prophase. There is a big nucleolus, and I could see that these chromosomes moved toward one another and began their synapses. But they were tiny, tiny little chromosomes. After they synapsed, they began to elongate—about fifty times what they were. As they elongated they got fatter, but they looked like railroad tracks. At this very elongated stage, I could determine chromomere patterns and so forth. Then they went into a diffuse stage [diplotene], which I felt sure was the stage when crossing over was occurring. But I couldn't see anything well with the light microscope. They came out of it suddenly and went into diakinesis and a metaphase of the first meiotic division very rapidly after this long period of being in diplotene. Then from that state on, from the stage of actual fusion of the two nuclei with the chromosomes intact, they remain in the chromosomes stage. They never leave being chromosomes; they are always chromosomes. They go into an anaphase in the first division as chromosomes and they just unravel, with their arms sticking down. Then they go into the second division. Perfectly normal second meiotic division. They go into an anaphase and the chromosomes elongate —the arms are hugely long. Very long arms coming down. All the time the genes in these chromosomes are active, because the *ascus* is growing bigger and bigger all the time, and you can see the activity is going on as this *ascus* grows. Then the chromosomes come down to the metaphase plane and form another division, the third division; [now] there are these eight nuclei. The eight nuclei then produce, during this stage, big plaques that all have a lot of microtubules. They develop at each pole at anaphase. When the nucleus membrane forms, [the nuclei] migrate along a line in the ascus. They all come down and orient themselves by this plaque, [spaced equally] along the ascus. Then from this plaque comes a set of fibers that surround each spore. They're still in the chromosome stage. Finally, they go into a division in the spore, and then you find nuclei again. . . . Well, that's the story, basically."

Seven days after coming out from under the eucalyptus trees, she gave a seminar on the meiotic cycle of *Neurospora.* In addition to the five days of actual work, many years of experience went into those observations. But above all, she felt it was "what happened under the eucalyptus trees" that was crucial. She had brought about a change in herself that enabled her to see more clearly, "reorienting" herself in such a way that she could immediately "integrate" what she saw.

That experience taught her an important lesson. "The point is that when these things happen—when you get desperate about something and you have to solve it—you do solve it and you know when you've solved it. You do something with yourself! You find out what's wrong, why you are failing—but you don't ask yourself that. I don't know what I asked myself; all I knew was that I had to go out under those eucalyptus trees and solve what it was that was causing me to fail."

Knowing that "everything was going to be all right," she found that, where before she had seen only disorder, now she could pick out the chromosomes easily. "I found that the more I worked with them the bigger and bigger [they] got, and when I was really working with them I wasn't outside, I was down there. I was part of the system. I was right down there with them, and everything got big. I even was able to see the internal parts of the chromosomes—actually everything was there. It surprised me because I actually felt as if I were right down there and these were my friends."

In telling this story McClintock sat poised on the edge of her chair, eager to explain her experience, to make herself understood; equally eager to avoid being misunderstood. She was talking about the deepest and most personal dimension of her experience as a scientist. A little later she spoke of the "real affection" one gets for the pieces that "go together": "As you look at these things, they become part of you. And you forget yourself. The main thing about it is you forget yourself."

A hundred years ago, Ralph Waldo Emerson wrote: "I become a transparent eyeball; I am nothing; I see all." McClintock says it more simply: "I'm not there!" The self-conscious "I" simply disappears. Throughout history, artists and poets, lovers and mystics, have known and written about the "knowing" that comes from loss of self—from the state of subjective fusion with the object of knowledge. Scientists have known it, too. Einstein once wrote: "The state of feeling which makes one capable of such achievements is akin to that of the religious worshipper or of one who is in love."[4] Scientists often pride themselves on their capacities to distance subject from object, but much of their richest lore comes from a joining of one to the other, from a turning of object into subject.

For McClintock, her experience with the *Neurospora* chromosomes provided confirmation of something she had known for many years; it focused the more diffuse feelings that had been with her since childhood. Twenty-five years earlier, her capacity for total absorption—her wish to be "free of the body"—had caused her to forget her own name; now it was something she seemed to have mastered. She had learned to summon it when needed, and to use it in the service of scientific discovery. She returned to Cold Spring Harbor ready to embark on the work that would lead to the major discovery of her career.

Without doubt, the year 1944 was a pivotal one in McClintock's professional life. She had always known her worth as a scientist, but now, with her election to the National Academy, she was granted the public recognition that corresponded to her own private evaluation. And with her growing sense of mastery of her own abilities, that private evaluation was doubly confirmed. Finally, after years of struggle, public and private judgments began to converge. She was now forty-two years old and approaching the peak of her career. Before the year was out, she was elected President of the Genetics Society of America—a position no woman had yet held. Returning to Cold Spring Harbor in the winter of 1944–1945, McClintock began the work that ultimately led to transposition. From a personal point of view, the timing could not have been better. Reinforced by her recent success, she had the confidence she would need to meet the most difficult challenge of her career.

It would be some years before the work McClintock began that winter crystallized and many more years before it was granted historical significance, but other developments of this period also had to await recognition of their proper place in biological history. Even the publication in 1944 of Avery's discovery of the genetic potency of DNA—perhaps the major event in twentieth-century biology—went relatively unnoticed. In the mid-1940s the simple mechanics of inheritance that molecular biology would reveal were as yet unknown. The complex processes of regulation and control that McClintock was about to uncover were not yet even to be dreamed of.

CHAPTER 8

Transposition

Back at Cold Spring Harbor, the results of the summer's planting waited to be read. As part of her continuing investigations of the new mutations produced by the breakage-fusion-bridge cycle, McClintock had grown a culture of seedlings produced by self-pollination of plants in whose early development one or both chromosomes 9 had been newly broken. Depending on the particular kind of breakage cycle that had occurred in the parent plants, these young seedlings included a number of familiar variants of the basic green color of most young seedlings: they might be white, light green, or pale yellow. But unlike other mutants that had been studied in corn, these "mutations" were apparently unstable within the life of a single plant. In each mutant seedling, streaks or spots of color could be seen that didn't belong—patches of pale yellow or green in a white leaf, or patches of normal green in a light-green or yellow leaf. The genetic instability that these patches reflected had been described in other organisms under the names of mutable genes, variegation, or mosaicism; but only rarely had such variation been seen in corn. In this crop "mutable genes" seemed to be everywhere.

Each patch of color defined a family of cells that had grown by division from a single (mutated) cell. Earlier mutations read as large patches; more recent mutations had produced a smaller progeny, showing as smaller patches. It followed that the number of patches of a given size provided a measure of the frequency of mutation at a given stage in the seedling's development. Many large patches would indicate that there had been a high rate of mutation early, and so on. From the distribution of patches, one could directly read off the history of genetic events that had accompanied the plant's development.

Examining this genetic "time chart," McClintock saw that each seedling exhibited a characteristic rate of mutation, unchanging within the life cycle of a given plant. A plant that started out with only a few mutating cells would maintain that habit throughout its life: just a few large patches, more small ones, but representing the same frequency relative to the pool of available cells at each stage. These were mutations that did not strike capriciously within the life of the single plant; whatever set them off was a factor that was constant. For McClintock, this regularity meant that something was controlling the rate of mutation.

Today, the concepts of regulation and control are part of the basic stock-in-trade of geneticists; in the 1940s, as McClintock recalls, among most geneticists "the idea of control was not even thought of." Of course, anyone who looked at an organism could see that development from a single fertilized cell is regulated: the fact that corn kernels yield corn plants is not fully explained by their starting out with the appropriate outfit of chromosomes. Cells must differentiate as they multiply if they are to produce the various kinds of tissue that make up the characteristic forms of the organism. But at that time geneticists were fully occupied with heredity, with the vicissitudes of the genome. The process by which cells differentiate and fulfill the specifications of their genotype was regarded as belonging to the remote domain of embryology.

McClintock's constant rates of mutation provided an instance of developmental regularity, directly linked to genetic events. The mutations acted as a tracer, allowing a history of cell differentiation to be read off; and that history turned out not to be random. She knew she was "onto something very important." For her, the question of how gene action might be regulated had always been a pressing one. "It seemed to me that if you look at the overall organism and how it develops, these things we call genes just *had* to be controlled." Her long familiarity with the outward life cycle of the corn plant and with the replication cycles of its chromosomes had trained her to see and think in terms of process. Now she was certain that she had a clue to the control of genetic events in normal development.

When scientists set out to understand a new principle of order, one of the first things they do is look for events that disturb that order. Almost invariably it is in the exception that they discover the rule. As McClintock continued to study these stable patterns of instability, she found cases that were exceptions within the exception. Occasional sectors of variegated tissue showed a rate of mutation different from that of the plant as a whole. Each of these distinct sectors presumably arose from an individual cell, and in a number of cases they seemed to arise in pairs. She sensed that here was the clue she needed and immediately "dropped everything else" to pursue it. The description of this discovery is presented in sober language in the annual report she wrote the next winter for the Carnegie Institution:

> . . . two distinctly delimited but adjacent sectors showed inverse relations in their mutation rates. In one sector the rate of mutation, as expressed by the number of green streaks in the leaf tissue, was greatly increased, whereas in the sister sector the number of green streaks was greatly reduced. The position of these twin sectors in the stalk and leaf suggested that they arose from two sister cells of the growing point.[1]

Her excitement over the recognition and the meaning it conveyed to her is more evident in her verbal account. "There would be a change during development such that there would be two sister cells formed that gave rise to two adjacent sectors, side by side, in which the pattern of gene action that you would see expressing itself was very different from the pattern you'd started with. It was also different in the two sister cells. In one it was very much increased in frequency, and the other very much reduced. . . . *Something* had to have occurred, at an early mitosis, to give such a different pattern. It was so striking that I dropped everything, without knowing—but I felt sure that I would be able to find out what it was that one cell gained and the other cell lost, because that was what it looked like. And I couldn't get it out of my head that one cell gained what the other cell lost, and that I would be able to find it out. . . . I don't know why, but I *knew* I would find the answer." That cryptic insight—"one cell gained what the other lost"—seemed to McClintock to be the next step toward an explanation of regulation.

She already had evidence of regular change in cells (the constant rate of mutation); now she was looking at a point in the plant's cell history where the rate of mutation was reset. A parent cell with one rate divided into two daughter cells with two new rates—one higher and the other lower than that of the parent. Here was the point of origin of the shift; from it she could hope to learn more about what controlled the rate. But even more important, here was an event that led to the differentiation of two cells—possibly just the kind of event needed to generate the different kinds of tissues of an organism. From this starting point she could begin to envision a science of developmental genetics.

The question of how each organism arrives at its own form was an integral part of her deepest concerns as a biologist. Cytogeneticists were trained to read the kind of evidence she had before her; but only a cytogeneticist with her particular constellation of interests would have seen its potential meaning. McClintock had studied embryology, and she knew about determination events.

Determination events, she stresses repeatedly, were the key idea. She explains that these were events that were then very well known to the embryologists. They had the property of bringing about a change in a cell that would only show up in the progeny of that cell many cell generations later. Embryologists had demonstrated the occurrence of such events, but they could not say anything about what they were—other than the fact that they took place. Now, she had a clue as to their nature. "In any case, the determination event had occurred as a consequence of a mitosis such that one cell gained what the other cell lost."

It took her two years to begin to understand what that statement might mean. In the meantime, she didn't have too much to go on—other, that is, than a deep conviction. Years later, Evelyn Witkin—at the time a young bacterial geneticist on fellowship at Cold Spring Harbor—asked her how she could have worked for two years without knowing what was going to come out. "It never occurred to me that there was going to be any stumbling block. Not that I had the answer, but [I had] the joy of going at it. When you have that joy, you do the right experiments. You let the material tell you where to go, and it tells you at every step what the next has to be because you're integrating with an overall brand new pattern in mind. You're not following an old one; you are convinced of a new one. And you let everything you do focus on that. You can't help it, because it all integrates. There were no difficulties."

Two years later she knew that what she was observing was a form of controlled breakage (or dissociation) in the chromosome —her first glimpse of transposition. "I had the answer. I had the answer that there was a component right adjacent to a gene, and it responded [by dissociating itself] to a signal sent out from another element." She called this system (the first of several she found illustrating such a mechanism of control) the *Ds-Ac* system: *Ds* for dissociator and *Ac* for activator.

But how she arrived at her conclusion is a story in and of itself. With each generation of corn plants studied, more and more anomalous data accumulated—data that could not be fit into any conventional framework. The ordering of these data required a new and elaborate theoretical edifice of which transposition was but one part. In all, it would be six years before the story was sufficiently complete to present to the scientific public, but much of the step-by-step evolution of her interpretation can be followed from the annual reports she wrote for the Carnegie Institution of Washington.

From these reports one can get a sense of the unfolding of her theory as a hierarchy of hypotheses, each more abstract and further removed from the objects of perception than the one before, yet, in concert, providing an internal logic so compelling as to give anyone who grasps that logic the sense of being able to "see" the abstractions themselves. Witkin, who worked in the same building and who followed McClintock's work day by day, came to feel that she, too, could "actually see genes turning on and off." It was easy for McClintock herself to lose sight of the difference between what could be seen with the relatively uneducated eye and what could be seen only with the help of the long chain of logical inference that, to her, had become second nature. In fact, a prodigious amount of cognitive processing intervened between the spots of pigment she could actually see on the corn plant and the controlling elements she ultimately came to write about. To invoke her own analogy, her "computer" was working full time—mediating between the spots, the patterns they formed, and her internal vision. The journey from the first clues to the final interpretation was a long one (it took six years), and the route was not straightforward. Indeed, until a degree of relative simplicity and order could be introduced by a new organizing scheme, complexity and confusion tended to grow rather than diminish. But, there was always a direction in which she was headed. The more complex and confusing the data grew, the more essential it was to have a point of reference to remind her of that direction, and sufficient ballast to keep her on course. Her point of reference was provided by her inner vision and the ballast by her extraordinary confidence. In the section that follows, the stages of her journey are sketched out in rough outline. It is not an easy journey to follow, but the attempt should result in a greater appreciation of what she had to do.

• • •

Transposition is a two-part process, involving the release of a chromosomal element from its original position and its insertion into a new position. The first hint McClintock encountered of a regulatory system involving such a mechanism appeared in a phenomenon that was itself, as it turned out, a consequence of the first part of the transposition process (i.e., of the release of a chromosomal element). In a number of plants expressing genetic instability, she found evidence of regularly occurring and highly specific breaks in the chromosome—a phenomenon she referred to as dissociation. The correlation of dissociation with genetic instability came early on, at a time when transposition was still far off.

Among the plants in that very first crop, there was one that was particularly noteworthy. Following self-pollination, a few of the kernels on this plant displayed a pattern of variegation so unusual they could not fail to catch the eye. Given their genetic constitution, these kernels should have been colorless. Instead, they contained well-defined sectors in which distinctive patterns of pigmentation could be seen—spots of color indicating the loss, in certain cells, of the dominant genetic factor responsible for the inhibition of (aleurone) color. Normally, that genetic factor (*I*) would be located in the short arm of chromosome 9; in these particular plants, the *I* factor should have been present on one of the two chromosomes 9 in each cell. The patterns in each sector were characterized by a uniform distribution of colored areas of essentially equal size; sectors differed by the frequency and size of such spots. The regularity of these patterns suggested that the *I* factor "had been systematically eliminated from some cells and that, in each sector, this had occurred at a particular rate and at a particular stage in the development of the endosperm tissue."[2] The key word here is "systematically." Loss of genetic markers had been observed before. What was new and noteworthy was the fact that the timing and frequency of the loss of this marker seemed to be regular, hence under some sort of control or regulation. Might the source of control itself be a genetic factor?

To try to find out what was going on, she planted the young seedlings grown from these kernels the following summer (1945), watching them as they grew. The major source of information would come from the offspring these plants would generate. Genetic analysis consists primarily of observing the passage of traits down through the generations. To find out if a genetic factor was responsible for these "systematic" losses, the first task was to determine the precise genetic constitution of these kernels (and the plants they gave rise to). A number of known genetic markers had been present in the chromosomes of the original parent plant from which these kernels had emerged, but as a result of the breakage-fusion-bridge cycle occurring in the early development stages of the parent plant and because of possible crossing over occurring during meiosis, the genetic sequence in the chromosomes of these particular kernels had to be newly determined. If, indeed, a genetic factor responsible for regulating the time and frequency of loss of the *I* factor did exist, it might then be possible, provided one had a map sufficiently well marked by known landmarks, to locate, or "map," that factor.

"Mapping" a known genetic marker with a clear and unambiguous phenotypic expression is a reasonably straightforward process. It requires a series of "crosses" or matings between pairs of parents of different genetic makeup. Assuming that the genes are laid out in a linear array on the chromosome, the probability of a genetic exchange (or crossover) between two chromosomes carrying different alleles of two distinct genetic markers is taken to be proportional to the actual distance between the markers. The reasoning is that the greater the distance between markers, the greater the likelihood of a break in that stretch of the chromosome, and hence of a crossover between chromosomes. The physical location of one factor relative to another is thus determined by the frequency of offspring that combine the genetic traits of both parents. The task of "mapping" an unknown genetic factor proceeds in the same way but can be considerably more complex, especially when the phenotype expression of that factor is as indirect as the one being tracked here. (McClintock was attempting to locate an element whose existence was only suspected—and which, if it did exist, expressed itself through the loss of certain other genetic factors.) Later, she would have to "map" factors at even greater remove from any visible indicator. But what looks hopelessly abstract, indirect, and circuitous to an outsider can, with enough experience, come to seem concrete and straightforward. The sequence of steps was simple to McClintock: pollen from the plant being analyzed was streaked on the silks of a variety of other plants of known genetic composition—selected to facilitate observation of the presence of the marker in question. The ensuing progeny were then analyzed. Some of the original plant's properties could first be observed in the mature kernel, and others, in the plant subsequently grown from that kernel.

After many such crosses, the precise genetic composition of the chromosomes undergoing dissociation was determined and the factor apparently responsible for the dissociation itself pinpointed. It lay on the short arm of chromosome 9, approximately one-third the way down from the centromere. Independent cytological analysis confirmed the fact that breakage was in fact occurring at the same point (which she ultimately called the *Ds* locus). In one nucleus she was actually able to see both chromosomes 9 broken at the same point, with the two sets of chromosome fragments in apparently perfect homologous association. By the summer of 1946, she could conclude that "breakage itself [in these plants] corresponds to the 'gene' mutations observed in other variegations."[3] Though she could not yet say what caused the "mutations" in the other variegation patterns, she suggested that a common mechanism—still unknown—underlay the control of the timing and frequency of all such "mutations."

A year later, the analysis was much advanced. Her earlier conclusion, now stated more clearly, stood:

> In this one case, mutability is expressed not by a visible phenotypic change in the action of a gene, but rather by dissociation of the bonds that normally would maintain a linear cohesiveness of this locus with an adjacent locus in the chromosome. As an ultimate consequence of the mutation, the chromosome is dissociated into two completely detached segments.[4]

But now, after following the pattern by which the genetic element responsible for dissociation sorted itself out in crosses with other genotypes, she was forced to conclude that not one, but *two* distinct genetic loci were involved: "Accumulating evidence indicated that the *Ds* locus will undergo dissociation mutations only when a particular dominant factor is present. This factor is designated *Ac* because it activates *Ds.*"[5] Something in *Ds* triggers dissociation, but something in *Ac* triggers *Ds.* Still more genetic crosses showed that *Ac* was located on the long arm of chromosome 9—far from the locus of *Ds.*

By this time, she had developed new staining techniques that permitted much sharper resolution of the chromosomes. She had also, by an astute choice of genetic crosses, managed to introduce cytological markers that would make it easy to distinguish those chromosomes that carried the *Ds* marker from those that did not. (*Ds* itself could not be seen under the microscope, but other markers, such as the small terminal knobs she had introduced in the vicinity of *Ds,* could.) Now she was in a position to push ahead with the cytological analysis: she could keep close tabs on the actual physical changes occurring (or not occurring) in the chromosomes that had (or did not have) the *Ds* locus. What she saw confirmed her earlier conclusions, but at the same time added new questions. She was still far from having an explanation of her original observations (dissociation alone would not suffice), let alone of the new data that grew correspondingly more complex.

And she still could not answer her central question: What caused the abrupt changes in the patterns of dissociation "mutations" in those sectors that had first captured her interest? It may have been obvious that these particular sectors had arisen

Barbara McClintock at work at Cold Spring Harbor, New York, 1947. (Permission of Marjorie M. Bhavnani.)

from an ancestral cell that had undergone some sort of change —she called it a "change of state" (presumably of the *Ds* locus) —but how? Unlike the actual dissociation mutation itself, this "change of state" appeared to be reversible. Once lost from a cell, a genetic factor could not return; thus, the progeny of cells that were pigmented because of the loss of a gene that was needed to inhibit pigment formation would remain pigmented. On the other hand, it was clear that the *patterns* of colored spots (the frequency and timing of dissociation) could change— and change back again.

By 1948, she knew that such changes of state also occurred

in the *Ac* locus—an increasingly remarkable, and powerful, locus. To complicate matters even more, new mutable loci—also related to *Ac*—were cropping up continuously. By this time, four other such loci had been identified on chromosome 9 alone; all four were controlled by *Ac* and all four exhibited such "changes of state." Her attention shifted to the study of the *Ac* locus itself. What kind of control does *Ac* exert over these other loci? And how does it express itself?

The *Ac* locus has no direct phenotypic expression; it can be identified only by its action on *Ds.* Nevertheless, that was enough to permit the identification of its presence or absence in any given plant, or on any given chromosomes. It behaved in inheritance (genetic crosses again) as a single, independent, and dominant locus. However, unlike more familiar genetic loci, it was not simply "on or off" in its effect. It became evident that the frequency and time of mutations controlled by *Ac* were themselves functions of the "dosage" of *Ac*; that is, since cells in the endosperm tissue carry three sets of chromosomes, plants having one, two, or three "doses" of *Ac* could be compared. The conclusion of such comparisons was that "the higher the *Ac* dosage, the later the occurrence of *Ds* [or other *Ac*-controlled] mutations."[6] Thus an increased dose of *Ac* results in an apparently lowered frequency of mutations.

In those plants carrying only a single dose of *Ds* and *Ac,* not only is the frequency of mutation high, but there is also great variability in mutation *patterns.* Of special interest was the fact that such plants showed large numbers of sectors with changed rates of mutation. Their appearance led McClintock to write, in 1948: "One is impressed with the resemblance of the mutation patterns in these various sectors to the patterns that have been obtained by combining various dosages of *Ac.*"[7] In other words, it looked as if the different sectors corresponded to different doses of *Ac*!

After all the data were put together, only one interpretation seemed to fit. It required assuming that: "(1) the *Ac* locus is composed of a number of identical and probably linearly arranged units, and (2) changes in the number of units can take place at the locus during or after chromosome reduplication such that one chromatid gains units that the sister chromatid loses."[8] Finally, she had an answer to what it might have been that "one sister gained [and] the other one lost." In this case, it was a unit of *Ac*; its shift from one sister chromatid to the other was responsible for the observed change of state.

But the story is still not finished. Before the full significance of such an exchange could be understood, other questions would need to be addressed. By what means does *Ac* "control" mutability in *Ds*, or in the other mutable loci under its control? And how does the loss or addition of a unit of *Ac* effect a change of state?

It had already been established that *Ac* (by whatever means) induced breaks at the *Ds* locus and that these breaks were generally followed by fusion of the broken ends. Continuing genetic analysis revealed that both *Ds* and *Ac* could sometimes be found in positions different from those originally identified. These two facts together suggested transposition—a term and concept McClintock introduced publicly for the first time in 1948. By 1949 she was certain not only that transposition occurred, but that it provided the key to the multiplicity of mutable loci controlled by *Ac*:

> Continued study . . . has revealed a type of event involving the *Ds* locus that appears to be responsible for the origin and subsequent behavior of all *Ac*-controlled mutable loci. This event brings about a transposition of the *Ds* locus from one location in the chromosome complement to another. In its new position, *Ds* responds to *Ac* just as it did in its previous position.[9]

In all such cases, the presence of *Ac* was mandatory. Her picture was as follows.

If two breaks are induced on opposite sides of the *Ds* locus, then a fragment of chromosome containing *Ds* is released and available for insertion or fusion at any other part of the chromosome complement where a concurrent break might have occurred. By such a mechanism, *Ds* could change its position. *Ac* "controls" the occurrence of such events by inducing the breaks at the original *Ds* locus. Although *Ds* is itself not visible, "its detection in the new position is easy . . . because it behaves as it did in its former position."[10] Namely, it undergoes subsequent dissociation.

Under the new interpretation, all other *Ac*-controlled mutable loci were nothing more than new sites of the *Ds* locus. If the new sites happened to be at functional genes, the presence of *Ds* would inhibit the normal functioning of that gene; normal functioning could resume only with the release of *Ds*. Transposition of *Ds* to the site of a functional gene was a particularly fortunate occurrence for the geneticist, because it provided a means of directly detecting the presence or absence of *Ds*. No longer need one rely solely on dissociation itself for detection of *Ds*; now one could infer its presence or absence from the inhibition or restoration of the action of a functional gene that, by definition, had direct phenotypic expression.

Finally, changes in state of the *Ds* locus could be attributed to the loss or addition of subunits of *Ds*. Just as breaks at the *Ds* locus could result in the release of the entire locus, so, at other times, they could result in the loss (or addition) of parts of the *Ds* segment. The hypothesis of transposition worked wonders: where the data had come to seem almost hopelessly complex, now they began to fall into an orderly and relatively simple scheme. But even now, the analysis was not complete.

The time and rate of occurrence of breaks in the *Ds* locus—whether they led to transposition, loss of chromosome fragments, or changes of state—are dependent on the dose of *Ac* present; such breaks never occur in the absence of *Ac*. It is in this sense that *Ds* is said to be under the control of *Ac*. But *Ac* itself was observed to undergo breaks, changes of state, and transposition. What controls these events?

The answer is *Ac* itself. In the paper she presented to the 1951 Cold Spring Harbor Symposium, McClintock wrote: "With any particular state or dose of *Ac*, the time and occurrence of changes of *Ac* are controlled by *Ac* itself."[11] Such feedback implies a built-in autocatalytic mechanism that, coupled with the segregation of chromosomal elements that occurs during meiosis and mitosis, has clear implications for cellular differentiation. The consequences of a particular change in *Ac* depend critically on the developmental stage of the cell in which that change occurs. In particular, transposition events occurring in the early gametophytic development of the plant (before fertilization) lead to dramatic differences among the kernels of a particular ear, whereas late-occurring events (during endosperm development) lead to a uniformity among the kernels, with the possibility of differential development of particular cell lines (or lineages) within each kernel. "If, with a particular *Ac* state, the time of such changes is delayed until late in the development of the endosperm, then all kernels should show this same late timing."[12] But, once again, not *all* the kernels do. The presence of occasional aberrant kernels on these ears suggests, in addition, that internal or external environmental alterations may cause an unusually early change in *Ac*. If so, changes in *Ac* both depend on the cellular (or nuclear) environment in which they occur and, in turn, generate different nuclear environments (and hence genetic fates) for the cell lines they give rise to. In any case, the mere dependency of changes in the state or location of *Ac* on the state of *Ac* in a given cell constitutes an explanation of the great variety of patterns of mutability observed; it provides a mechanism for differentiating kernels, cell lines, or individual cells, depending on the stage at which such changes occur.

But what does all this say about normal development? Weren't these mutable loci notable by virtue of their very abnormality? McClintock thought not. The variegation patterns that had led her down this long and arduous road were not, as some might think, expressions of a breakdown of normal processes. Rather, they could be seen as "merely an example of the usual process of differentiation that takes place at an abnormal time in development."[13] Viewed in this way, they provide a means of understanding the contribution of the nucleus, or even the cell as a whole, to the control of the differentiation process. The potential for particular types of genetic action varies with differences in the nuclear composition of different cells. Such differences can result from transposition of controlling elements between sister chromatids, thereby giving rise to sister nuclei that are no longer alike, or they might arise in other ways. The events producing such nonequivalence just might be the long-sought-after "determination events" of the embryologist. But whichever way they arise, they demonstrate that the key to understanding development is the recognition that, rather than genes per se, "it is organized systems that function as units at any one time in development."[14]

• • •

Six years of intensive labor enabled McClintock to complete her early vision with a fully articulated and abundantly supported theoretical structure. Reams of data filled her office cabinets: one card for every genetic cross, fat books recording the data from these crosses, extensive tables compiling the data, accompanied by interpretations, and a three-part voluminous manuscript just on transposition alone. "Transposition was absolutely nonsensical to biologists then," she remembers. It was, therefore, essential that her data anticipate every possible objection. "I went overboard collecting evidence showing it had to occur—until there was absolutely no doubt." It was far too much to publish. She had written the manuscript on *Ds* transposition in sequential installments—as private monograph-length communications to Marcus Rhoades, who by that time had moved to the University of Illinois. She sent a carbon copy to Stanley Stephens, a bright and sympathetic geneticist at North Carolina State University in Raleigh who worked on cotton. At Cold Spring Harbor, she discussed her results daily with Evelyn Witkin.

Evelyn Witkin, a student of the geneticist Theodosius Dobzhansky, came to Cold Spring Harbor for the first time in the summer of 1944; the following summer she came to stay for ten years. Today, Professor Witkin holds the Barbara McClintock Chair at Rutgers University—a chair she named as an acknowledgment of the great debt she feels to McClintock for many years of inspiration. From the beginning, the older woman engaged the younger one's affection and admiration, and Evelyn Witkin quickly became someone with whom McClintock could share her intuitions and discoveries. She guided her through the intricacies of maize genetics and explained the meanings of the data that emerged.

Witkin was not herself a maize geneticist, but, under McClintock's tutelage, she soon became competent at reading the patterns so well that she, too, could "actually see genes turning on and off at definite times."[15] Her own lab was in the same building, and when something new appeared, McClintock would call her up, and she would run down to look. Learning maize genetics—so different from and so much more complex than the bacteria she worked on—took a lot of time. But the excitement of being able to follow the course of McClintock's search made it well worth her while.

"Just looking over her shoulder, looking at the spots, you could visualize what was going on—she made you see it," Witkin says. "She was even able to convey it to someone who was completely outside the field. She was able to make it real."

Witkin recalls that at the time, the problem of differentiation was on everyone's mind—the question of how one could account for the differential development of tissues generated by the same germ cells had even been put to her on her doctoral exam—but no one had any answers. What McClintock was finding was "completely unrelated to anything we knew, it was like looking into the twenty-first century."

However, not many others shared her confidence. McClintock recalls that "Evelyn was the only one who really had any understanding of what I was doing." Marcus Rhoades was certainly supportive—he had absolute confidence in her—but she wasn't sure he quite understood. Most others at Cold Spring Harbor didn't know enough maize genetics, and, until 1950, the only generally available account was in the annual reports she wrote for the Carnegie Institution of Washington. Compared to the papers that were to come later, these reports make remarkably easy reading, but at the time few people read them. They were not widely circulated, and they presented almost none of the supporting data. Scientists do not normally look to annual reports of an institution to learn what is happening in their field.

In the fall of 1950, McClintock published a brief account of the transposition of *Ds* and *Ac* in the *Proceedings of the National Academy of Sciences.* She introduced her topic under the title "The Origin and Behavior of Mutable Loci in Maize," by calling attention to the parallels between the system she was studying and other kinds of genetic instability (including the notorious "position effect") that had been studied in *Drosophila,* while emphasizing the similarity in the mechanism underlying variegation in all such organisms. But the obvious occasion on which to present the full account of this work, with all of its radical implications, was at the next annual Cold Spring Harbor Symposium. As the time grew near, she became more and more apprehensive. How was she going to be able to fit in enough of the pieces in the time allotted to her to make it coherent? Would she be able to communicate her "understanding" to others? She was sufficiently aware of the disparity already present between her own thinking and that of her colleagues to know that many of them would have difficulty seeing the implication of her new findings.

CHAPTER 9

A Different Language

McClintock was right to be apprehensive. Her talk at the Cold Spring Harbor Symposium that summer was met with stony silence. With one or two exceptions, no one understood. Afterward, there was mumbling—even some snickering—and outright complaints. It was impossible to understand. What was this woman up to?

Somehow, she had "missed." Her effort to explain the logic of her system seemed to have failed utterly. Of course she would try again—it is a commonplace for talks to misfire—but the disappointment must have been enormous. She had unveiled her creation, a beautiful explanatory model with full supporting evidence, the object of six years of loving attention and grueling hard work, and her colleagues had turned their backs.

Apprehension has at least the function of preparing us for adversity. But McClintock could hardly have been prepared for the extent of her failure. She tried again—in a few seminars, in a detailed account of the work (including much of the data that earlier brief accounts had not included), which she published in *Genetics.*[1] Five years later she gave another presentation at the 1956 Cold Spring Harbor Symposium. In the interim, the mechanisms of control and regulation she had unravelled had shown themselves to be even more complex, so that her work was now more, rather than less, difficult to follow—and the climate around her had grown even less receptive than it had been in 1951. She received only two reprint requests for the *Genetics* article (published in 1953), and, if anything, there was more shoulder shrugging at the 1956 presentation than there had been in 1951.

No one is ever truly prepared for so bitter an anticlimax. But McClintock must have been especially ill prepared. She was hardly a novice in her field, and despite the many institutional difficulties she had experienced, she was accustomed to scientific success. Above all, she was accustomed to the respect and admiration of her colleagues. By 1951, she was one of the dignitaries of her field, and scientists of her stature do not expect their work to be rebuffed out of hand.

Individuals who have as strong a conviction of being right as McClintock had expected to be listened to. In the long run, that conviction served her extremely well; it provided the protection she needed in order to continue her work. But in the short run, it could only have exacerbated the shock. She tends to make light of it now, but at the same time she admits that the 1951 symposium "really knocked" her. "It was just a surprise that I couldn't communicate; it was a surprise that I was being ridiculed, or being told that I was really mad. And it required a readjustment." In the long run, she claims, it was good for her, but some of those who were closest to her feel that it was a readjustment with a high personal cost. "Later on, there were years in which I couldn't talk to anybody about this and I wasn't invited to give seminars either." She remembers one well-known geneticist, a frequent visitor to Cold Spring Harbor, saying, " 'Now, I don't want to hear a thing about what you're doing. It may be interesting, but I understand it's kind of mad,' —or words to that effect." A few were less polite. She repeats as a "funny story"—told to her by her friends lest she should hear it accidentally—that a prominent geneticist called her "just an old bag who'd been hanging around Cold Spring Harbor for years."

Of course, not all her colleagues were as dismissive. She had good friends, a few staunch allies (Evelyn Witkin was unshakable), and a number of loyal admirers scattered around the country. A few corn geneticists knew enough to appreciate what she was doing, and they came to talk to her, to exchange technical information, and to pick up tips (or seeds). Prominent among these were Royal Brink and Peter Peterson, whose own work had led them in strikingly similar directions. In 1952, Brink (together with Robert A. Nilan) published evidence for transposition that was also inferred from the observation of twin sectors, and in 1954 he and P. C. Barclay were able to show that a controlling element they had isolated independently was an instance of the *Ds-Ac* system. Peter Peterson isolated a mutant in 1953, which by 1960 he was able to show was an instance of another system of regulation and control that McClintock had by then detailed: the *Spm* system.[3] Indeed, the work of Brink and Peterson echoed and supported many of her own findings over the next two decades. "We were compatible because we were working on the [same] material." But even they were "on different levels of thinking—that's where we couldn't communicate."

But the presence of a few friends, allies, admirers, and even fellow workers could not outweigh the experience of rejection by the overwhelming majority of her colleagues. If McClintock had been physically isolated before—working alone, without the benefit of students, post-docs, or immediate colleagues—she had nonetheless always been able to maintain a dialogue with geneticists around the country. To be sure, her position was peripheral in a number of important ways. But, with increasing success and growing reputation, she had moved closer to the center of professional action than any other woman in her field. She attended frequent meetings, received frequent invitations to give seminars, and grew accustomed to visits from colleagues from all over the world. Even at Cold Spring Harbor, where her colleagues worked on different problems and different organisms, and where some of her relations with others seemed difficult to her, she was highly respected and her rapport with almost all of her colleagues was at least cordial.

Now all that seemed to change. As her isolation reached into her intellectual and professional lives, it deepened and took on new dimensions. Her efforts to talk about her work during the 1950s were in vain, and the principal consequence was that she stopped talking and, except for the annual reports in the *CIW Yearbook,* stopped publishing.

She withdrew further into her work, protected more and more by her "inner knowledge" that she was on the "right track," but at the same time becoming increasingly wary about confronting potentially hostile audiences, and even about visits from unsympathetic colleagues. Her lab remained open to anyone who genuinely wanted to listen or even just to talk, but she had always had a quick sharp tongue and now used it to protect herself whenever she felt the need. Lotte Auerbach, an animal geneticist from the University of Edinburgh, was one of those who wanted to listen. She found McClintock surprisingly patient and clear. In the space of a single afternoon McClintock was able to explain her work in sufficient detail to leave Auerbach not only convinced, but enormously impressed—and enthusiastic enough to try (unsuccessfully as it turned out) to convince others upon her return to Europe. But Auerbach also remembers Joshua Lederberg returning from a visit to McClintock's lab with the remark: "By God, that woman is either crazy or a genius." As Auerbach tells it, McClintock had thrown Lederberg and his colleagues out after half an hour "because of their arrogance. She was intolerant of arrogance. . . . She felt she had crossed a desert alone and no one had followed her."[3]

McClintock's own account puts a more cheerful face on those painful years. Once she got used to the realization that most geneticists did not want to know what she was working on, "it didn't bother me at all." Later, she says, she came to feel actually "glad." "It was fortunate, because people love to talk about themselves and their work, and I had an opportunity to listen. And I listened very carefully." One reason she found it worthwhile was that so much was going on in genetics during those years. "I was being educated, and it was an opportunity for me I do not regret; in fact, I think it was a great opportunity not to be listened to, but to listen. Difficult as it may seem."

But for all the new things there were to learn about, and for all the pleasure she got from her own work, the atmosphere at Cold Spring Harbor was sufficiently uncongenial for her to consider leaving. More than once she wrote to Marcus Rhoades for help in finding another place. She never did leave, but to this day she will not give a seminar at her own institution.

On the face of it, there is no good reason to think that things would have been any better elsewhere. But Cold Spring Harbor does hold a rather special place in the biological history of that period, and some of the features that helped to make the 1950s at Cold Spring Harbor unique may have widened the gap that separated McClintock from the rest of her scientific community. To see why that might have been, we need to face the question of what went wrong between McClintock and her colleagues.

One cannot explain the initial responses of her community (she was "incomprehensible," "mystical," even "mad") by saying, "Well, she was wrong." The notion that she was "ahead of her time" is more apt, but does not really help. Does it mean that her hunches were lucky and that one needed to wait three decades for adequate proof? Or that the evidence she had amassed could not be fitted into the expectations and preconceptions of the time? Because of the later legitimization of transposition, we are both given the opportunity and, in fact, granted the mandate to discuss these questions.

When we say we do not "understand" someone's argument, we usually mean that we do not perceive the underlying logic that makes all the parts of what the other is saying go together; we don't "get the picture." To say someone is obscure is to put the burden of that failure in communication on the speaker, something we feel entitled to do when many or the majority of supposedly competent listeners do not understand. Clearly, this is a judgment that, although it puts the onus on the speaker (or writer), also reflects the particular presuppositions, experiences, and expectations of that majority. Obscurity is, even with the most sophisticated audience, a relational attribute; it refers not simply to the properties of what is spoken (or written) but to how the message is heard (or read) as well. Looking at it in this way, we can avoid the futile argument about whose "fault" that failure of communication was. To be sure, when a scientist has new results to present, we generally agree that it is the responsibility of that scientist to bridge the gap that has arisen between his or her own expertise and that of others. McClintock did not effectively fulfill that responsibility. The question is: Could she have?

Perhaps, in 1951, it would have been possible. But given her initial failure and the subsequent events that took over in biology, the gap between McClintock and her colleagues soon became too wide and, I suggest, unbridgeable. There seem to be two separate but not unrelated sources of difficulty: one pertaining to the revolutionary implications of her findings, and the other to the particular nature of her knowledge and understanding.

It is a commonplace about scientific discourse that the more a claim is at odds with accepted beliefs, the more resistance it encounters. (It is also the case that any divergent claim is by its nature hard to understand, even for those who listen with good will.) And the results McClintock reported in 1951 were totally at variance with the view of genetics that predominated. The biggest problem was, if genetic elements were subject to a system of regulation and control that involved their rearrangement, what meaning was then left to the notion of the gene as a fixed, unchanging unit of heredity? Central to neo-Darwinian theory was the premise that whatever genetic variation does occur is random, and McClintock reported genetic changes that are under the control of the organism. Such results just did not fit in the standard frame of analysis.

But it was not only the ideas themselves that were foreign, and hence difficult to grasp for most geneticists; the very kinds of evidence she presented, or rather the patterns it formed, were also difficult to follow. When McClintock initially presented her work at the 1951 Cold Spring Harbor Symposium, she undertook to describe a system she had lived and worked with for over six years—in a plant she had worked with for almost thirty years. Her knowledge of maize was more intimate and more thorough than that of anyone else in the audience. Furthermore, while she had always been accustomed to working largely alone, the six years of work on this particular system had proceeded in special isolation. True, she had talked almost daily with Witkin and had corresponded regularly with Rhoades (and Stephens), but that communication had been, for the most part, one way. She had developed her ideas alone, without benefit of the mutual understanding that can grow out of ongoing discussion with colleagues. Once the details had fallen into place and the patterns had become clear to her, she had to confront the task of making it visible to others. To accomplish this task, she needed a language of discourse in common with her audience.

Scientists and philosophers of science tend to speak as if "scientific language" were intrinsically precise, as if those who use it must understand one another's meaning, even if they disagree. But, in fact, scientific language is not as different from ordinary language as is commonly believed; it, too, is subject to imprecision and ambiguity and hence to imperfect understanding. Moreover, new theories (or arguments) are rarely, if ever, constructed by way of clear-cut steps of induction, deduction, and verification (or falsification). Neither are they defended, rejected, or accepted in so straightforward a manner. In practice, scientists combine the rules of scientific methodology with a generous admixture of intuition, aesthetics, and philosophical commitment. The importance of what are sometimes called extrarational or extralogical components of thought in the *discovery* of a new principle or law is generally acknowledged. (We may recall Einstein's description: "To these elementary laws there leads no logical path, but only intuition, supported by being sympathetically in touch with experience."[4]) But the role of these extralogical components in persuasuasion and acceptance (in making an argument convincing) is less frequently discussed, partly because they are less visible.[5] The ways in which the credibility or effectiveness of an argument depends on a realm of common experiences, on extensive practice in communicating those experiences in a common language, are hard to see precisely because such commonalities are taken for granted. Only when we step out of such a "consensual domain"—when we can stand out on the periphery of a community with a common language—do we begin to become aware of the unarticulated premises, mutual understandings, and assumed practices of the group.

Even in those subjects that lend themselves most readily to quantification, discourse depends heavily on conventions and interpretation—conventions that are acquired over years of practice and participation in a community. Thus even an argument in theoretical physics—the most extensively mathematicized of all sciences—depends on communally shared conceptions of the meaning of the terms in an equation and on the relation of the equations to the processes they represent. When such a common "language" is absent, arguments cease to be effective, even if the equations are impeccable.

An especially good illustration of the role of such different "languages" in physics is provided in Freeman Dyson's recent autobiographical reminiscences. Dyson recalls his own role as "interpreter" of the brilliant but poorly understood Richard Feynman. Seeing Feynman unable to communicate with Hans Bethe and the other grand masters, Dyson decided that his job "would be to understand Dick [Feynman] and explain his ideas in a language that the rest of the world could understand."[6] Obviously, Feynman's own attempts to explain his simple but unorthodox methods weren't succeeding. "Nobody understood a word Dick said."[7] And when Dyson finally "got the hang" of Feynman's arguments, he could understand why.

> The reason Dick's physics was so hard for ordinary physicists to grasp was that he did not use equations. . . . Dick just wrote down the solutions out of his head without ever writing down the

> equations. He had a *physical picture* [my italics] of the way things happen, and the picture gave him the solutions directly with a minimum of calculation. It was no wonder that people who had spent their lives solving equations were baffled by him. Their minds were analytical; his was pictorial. My own training . . . had been analytical. But as I listened to Dick and stared at the strange diagrams that he drew on the blackboard I gradually absorbed some of his pictorial imagination and began to feel at home in his version of the universe. . . . [Over time, Feynman's ideas have been] slowly absorbed into the fabric of physics, so that now, after thirty years, it is difficult to remember why at the beginning we found it so hard to grasp. I had the enormous luck to be there at Cornell in 1948 when the idea was newborn, . . . I witnessed the concluding stages of the five-year-long intellectual struggle by which Dick fought his way through to his unifying vision.[8]

In fields less amenable to quantification and more dependent on qualitative judgment than physics, the weight of interpretive tradition is correspondingly greater. When divergences of experience and "language" arise in such fields, the presence of interpreters—translators such as Dyson—is critically important. Some mediation between disparate experiences and "languages" must be established. But by the same token, the task of such mediation can be much more difficult in fields more dependent on qualitative judgment. Perhaps the greatest difference among scientific fields lies in the kinds of practice necessary for effective communication. In theoretical physics much of that practice, and, in particular, much experience with the interplay between verbal and mathematical symbols, can be acquired by reading the printed page. In other fields, it requires more physical, active participation with the material itself. Cytogenetics is just such a field.

Arguments in cytogenetics employ an interplay of qualitative and quantitative reasoning in which the quantitative analysis rests on a host of prior judgments that remain necessarily qualitative. In particular, before the effects of specific genetic crosses can be counted, distinct phenotypic and cytological traits need to be identified. Both of these processes of identification require kinds of experience not easily communicated to those who have not participated in the actual observations. They require an extensive training of the eye. And McClintock's eye was surpassingly well trained. "Seeing," in fact, was at the center of her scientific experience.

For all of us, our concepts of the world build on what we see, as what we see builds on what we think. Where we know more, we see more. But for McClintock, this reciprocity between cognitive and visual seems always to have been more intimate than it is for most. As if without distinguishing between the two, she knew by seeing, and saw by knowing. Especially illustrative is the story she tells of how she came to see the *Neurospora* chromosomes. Unwilling to accept her failure to see these minute objects under the microscope—to pick them out as individuals with continuity—she retreated to sit, and meditate, beneath the eucalyptus trees. There she "worked on herself." When she felt she was ready, she returned to the microscope, and the chromosomes were now to be seen, not only by her, but, thereafter, by others as well.

If this were a story of insight arrived at by reflection, it would be more familiar. Its real force is as a story of eyesight, and of the continuity between mind and eye that made McClintock's work so distinctive and, at the same time, so difficult to communicate in ordinary language.

Through years of intense and systematic observation and interpretation (she called it "integrating what you saw"), McClintock had built a theoretical vision, a highly articulated image of the world within the cell. As she watched the corn plants grow, examined the patterns on the leaves and kernels, looked down the microscope at their chromosomal structure, she saw directly into that ordered world. The "Book of Nature" was to be read simultaneously by the eyes of the body and those of the mind. The spots McClintock saw on the kernels of corn were ciphers in a text that, because of her understanding of their genetic meaning, she could read directly. For her, the eyes of the body *were* the eyes of the mind. Ordinary language could not begin to convey the full structure of the reading that emerged.

Now that a number of biologists have become motivated to understand McClintock's work, one can get a glimmer of what the difficulties have been and what it takes to overcome them. Over and over again, these scientists locate their "breakthrough" in the experience of "seeing the patterns" on the actual kernels of corn; finding, in one biologist's words, "a single color photograph more illuminating than all her papers put together."

Evelyn Witkin's description of how her own understanding had developed, by looking over McClintock's shoulder, is further illustrative. Looking at the material, guided by McClintock's running narrative, she, too, learned to "actually see the genes turning on and off."

Witkin can be said to have learned a special kind of language —a language in which words and visual forms are woven together into a coherent structure of meaning. Once knowing the "language," Witkin could say that the arguments McClintock presented were as convincing and the proof as rigorous as any she knew. But to others with whom McClintock had not interacted as extensively, they were "incomprehensible."

In order to "see" what McClintock "saw," Witkin had to learn more than a new "language"; she needed to share in McClintock's internal vision. In that sense, "seeing" in science is not unlike "seeing" in art. Based on vision, our most public and our most private sense, it gives rise to a kind of knowledge that requires more than a shared practice to be communicable: it requires a shared subjectivity.

In his classic work *Art and Visual Perception,* Rudolf Arnheim reminds us:

> The human mind receives, shapes, and interprets its image of the outer world with all its conscious and unconscious powers, and the realm of the unconscious could never enter our experience without the reflection of perceivable things. There is no way of presenting one without the other.[9]

Inevitably, "seeing" entails a form of subjectivity, an act of imagination, a way of looking that is necessarily in part determined by some private perspective. Its results are never simple "facts," amenable to "objective" judgments, but facts or pictures that are dependent on the internal visions that generate them. In ordinary life, these private perspectives seldom emerge as discrepancies; the level of shared vision required for people to cooperate is usually met. But science and art alike make tougher demands on intersubjectivity: both are crucially dependent on internal visions, committed to conveying what the everyday eye cannot see.

The role that vision—in both the literal and the metaphoric sense—plays in scientific creativity has not been lost to others. Gerald Holton, an astute observer of the scientific imagination, has commented on its importance in the creative experience of two particular scientists, Robert Millikan and Albert Einstein. According to Holton, as Saint Thomas saw seraphim and Jean Perrin saw atoms, so Millikan saw electrons.[10] Holton cites three critical factors in Millikan's style of research:

> (1) His capacity of looking with fresh, clear eyes at what was going on; (2) his powers of visualization as an aid in drawing conclusions; (3) behind all these, almost unconfessed and certainly unanalyzed, a preconceived theory about electricity which gave him eyes with which to look and interpret.[11]

Better known is the example of Einstein, who was led to relativity by imagining the visual experience of a traveller on a beam of light. Later, he commented: "During all those years there was a feeling of direction, of going straight toward something concrete. It is, of course, very hard to express that feeling in words. . . . But I have it in a kind of survey, in a way visually."[12] For Einstein, mathematics itself was to be "seen." He wrote: "The objects with which geometry deals seemed to be of no different type than the objects of sensory perception 'which can be seen and touched.' "[13] And Holton comments: "The objects of the imagination were to him evidently persuasively real, visual materials, which he voluntarily and playfully could reproduce and combine, analogous perhaps to play with shapes in a jigsaw puzzle.[14]

In both of Holton's examples, as in Dyson's description of Feynman, the subjective aspects of vision were important primarily in the processes of discovery. The requirements of verification lay elsewhere. For McClintock, given the nature of the evidence with which she worked, discovery and verification both relied equally on "seeing." In this sense, her medium more closely resembles that of the artist: in both cases, the success of the end product depends on the possibility of a like-minded vision in the viewer. The very task of consensual validation, of "appealing to the evidence," requires a degree of intersubjectivity, of shared vision as well as shared language, on which she found she could not count. As a consequence, her reading of the "Book of Nature" remained her own.

• • •

Throughout the 1950s, the prospect of a common vision receded further and further. As McClintock's own research continued to enrich the picture she had formed in the 1940s, the specific events occurring elsewhere in genetics were leading most biologists down quite a different path. But that divergence of paths is another story, the telling of which requires us to look at what those new events were and what they implied.

CHAPTER 10

Molecular Biology

We now know that the 1950s was the decade of the molecular revolution in genetics. But—ironically or appropriately—the decade opened on a crisis in classical genetics in which, for a brief while, quite another sort of revolution appeared to be possible. With a number of geneticists voicing growing dissatisfaction with the classical concept of the gene, it was a time for renegades to prosper. In this climate, McClintock's work could have been seen, and was seen by some, as providing crucial support for a revolution that in fact did not occur.

The Cold Spring Harbor Symposium of 1951 on "Genes and Mutations" provides a vantage point from which to compare the new status of the gene with what it had been ten years earlier when a symposium had been held on "Genes and Chromosomes." At the same time, it provides a small glimpse into the decades to come. The difference in backward and forward perspectives is both notable and instructive.

As the organizer of the 1951 symposium, Milislav Demerec introduced the published proceedings with the remarks: "The original problem of defining the unit of heredity, which almost fifty years ago was designated 'the gene,' has not yet been solved. In fact, the large body of information accumulated since 1941 has made geneticists less certain than ever about the physical properties of genes."[1]

The symposium of 1941 had marked a decade of almost unbroken progress in our understanding of the chromosomal basis of genetics. The flowering of cytogenetics research during that period had yielded a wealth of evidence that bolstered geneticists' confidence in the physico-chemical basis of genes and, perhaps more to the point, in their integrity as *discrete* physical units. Demerec summed up the view of genes that prevailed in 1941 as follows: "Ten years ago they were visualized as fixed units with precise boundaries, strung along chromosomes like beads on a thread, very stable, and almost immune to external influences."[2] With the exception of Richard Goldschmidt and a few others, those who had initially resisted so reductionist a view, and who had been inclined to more holistic conceptions of genetic organization, had been won over. Even those more conservative thinkers who had cautioned that genes were no more than "useful abstractions" ("symbols," as McClintock called them) seemed to have been persuaded.

But in 1951, this conception of genes as "beads on a string," as discrete molecular entities, was showing signs of wear and tear. "Now," Demerec went on, "they are regarded as much more loosely defined parts of an aggregate, the chromosome, which in itself is a unit and reacts readily to certain changes in the environment." These remarks were prompted largely by three papers in the first session of the 1951 symposium, a session devoted to the "Theory of the Gene." Goldschmidt presented the first paper, Lewis Stadler the second, and McClintock the third. All three focused on the problem of mutation.

Both Goldschmidt and Stadler pointed out that everything known about genes was known from the study of mutations. The very concept of a discrete gene was inferred from the existence of localized mutations. But what were mutations?

Without knowing what the mutational event consisted of, what evidence was there, finally, to conclude that these events induced changes in a discrete quantity called the gene? With increasing evidence of mutagenic events that lead to large-scale chromosomal rearrangements, what was there to prevent one from concluding that *all* mutagenic events were chromosomal rearrangements? Goldschmidt argued that what appeared to be point mutations were simply the "effects of rearrangements on a submicroscopic scale." What need, then, was there for the hypothesis of a discrete gene? Having argued for many years that the gene was merely a figment of geneticists' imaginations, that the genetic unit was the chromosome and genetic traits were reflections of the serial arrangement of the chromosome, Goldschmidt felt that recent work (especially McClintock's) on mutable genes and chromosomal rearrangements vindicated his arguments. "I add with an understandable modicum of satisfaction that more and more workers in the field are beginning to direct their thoughts in a similar direction."[3] His interest in McClintock's work derived less from its implications for control and organization than from its demonstration of transposition —and especially from McClintock's conclusion (in her 1950 paper) that the phenotypic changes she observed resulted not from changes in the actual genes, but from "changes in a chromatin element other than that composing the genes themselves."[4] In Goldschmidt's reading, this work elevated the phenomenon of what was called the position effect to near universality. "Today," he said, "after McClintock's brilliant work, it has become one of the most common genetical phenomena in plants, if one is permitted to assume that all the numerous cases of so-called mutable genes have the same cytological basis."[5]

Goldschmidt's support, by itself, was perhaps more of a liability than an asset. Almost no one was willing to accept his extreme conclusion that "all mutants, whatever their frequency, place, cytological visibility as arrangements, or non-visibility are position effects,"[6] but his critique of the logic that had in the past led geneticists to attribute phenotypic changes to changes in a hypothetical "gene molecule" was clearly finding more support than it had in the past.

Stadler, more sober and a far more widely respected leader of the field, had become troubled by some of the same logical weaknesses that Goldschmidt attacked. Stadler was not opposed to the concept of the gene per se, but by 1951 his careful and probing theoretical analysis had led him to a critique of the gene theory that may have been, particularly given his known caution, more persuasive than Goldschmidt's. The arguments Stadler presented in 1951, just three years before his death, were essentially the same as those summed up in his final paper (the editors called it his "valediction"), published posthumously in *Science.* His central conclusion in both papers is that the assumption that point mutations are gene mutations is unwarranted: "Mutations in which the altered phenotype is produced by a gene mutation (that is, by the production of a new gene form) cannot be distinguished from . . . extragenic mutations by any positive criterion."[7] And since all our knowledge of genes depends on the occurrence of mutations, this means that our definition of the gene is left hanging in mid-air, with a built-in circularity. "To insist that [X-ray induced] mutants represent qualitative changes in specific genes because that is what we mean by gene mutation is to adopt the dictum of Humpty Dumpty in *Through the Looking-Glass,*" Stadler argued. " 'When I use a word,' Humpty Dumpty said, 'it means just what I choose it to mean—neither more nor less.' "[8]

Although Stadler was not prepared to dispense with the gene concept altogether, he felt compelled to acknowledge that, almost fifty years after its introduction, it remained essentially undefined. It was a hard conclusion to come to for the man who, along with Muller, had developed the technique of X-ray induced mutations as a means of elucidating the behavior of the genetic substance. Unlike Goldschmidt, who had made a profession of refuting the gene theory, his criticisms came after a lifelong career of attempting to do just what he now claimed had failed. What he felt was urgently needed was the development of a more adequate method of analysis that would enable geneticists to break out of the circularity in which they currently found themselves.

In the context of the issues both Goldschmidt and Stadler raised, McClintock's work on transposition was an extremely welcome contribution. It focused on mechanisms of "genetic" variations that, granted the assumption that genes were fixed and autonomous beads on a string, had to be attributed to "extragenic" chromosomal modifications. In so doing, it helped to expose the falsity of the assumption that "mutations" necessarily represent some transformation of the gene itself. Stadler wrote in 1954 (referring to both the 1950 and 1951 papers of McClintock): "The remarkable studies of McClintock on mutational behavior in maize . . . have shown the far-reaching importance of this limitation [of our assumptions] in the experimental study of gene mutation."[9]

Clearly, McClintock's demonstration of transposition as a mechanism for generating changes in genetic function had been heard (and at least partially understood) by some. In fact, by the early 1950s a climate had developed that ought to have been favorable to the reception of radically new ideas in genetics. Many others joined Stadler and Goldschmidt in their discontent with the classical concept of the gene as an inviolable and autonomous unit—simultaneously and inseparably a unit of function, of mutation, and of recombination. If Stadler found contradictions in his mutation studies, other workers found contradictions in recombination studies, which, as long as the gene was regarded as an indivisible unit, remained irresolvable. To all these workers, a conceptual change of one kind or another seemed inescapable. In this sense, McClintock's presentation in 1951 *was* well timed, and, had less been packed into the brief time available, or had her presentation been more clearly articulated, it might well have succeeded in capturing her audience's imagination.

But other movements were also afoot in 1951—movements that would soon drown out the criticisms of Goldschmidt and Stadler. Just a few sentences after remarking on the decline of confidence in the "gene" as a distinct unit, Demerec went on to observe:

> One of the most remarkable developments of these ten years concerns the organisms used in gene studies. In 1941 about thirty percent of the Symposium papers reported research carried on with *Drosophila,* and only six percent dealt with microorganisms; whereas this year only nine percent of the papers relate to *Drosophila,* and about seventy percent to microorganisms.[10]

The shift in focus from multicellular animals and plants such as *Drosophila* (and maize) to bacteria and bacteriophage was to become even more pronounced in the years to come. In calling attention to it, Demerec had touched upon one of the most telling indicators of the revolution that was, by 1951, already in the making.

If *Drosophila* had the advantage over maize of yielding a new generation every fourteen days, instead of annually, then bacteria were incomparably better. A bacterium divides in two every twenty minutes, and a bacteriophage undergoes several rounds of replication in half that time. On the other hand, the differences between bacteria and "higher" organisms are so vast as to suggest that the primary division of living organisms is not between plants and animals but rather between what have come to be called prokaryotes and eukaryotes. In prokaryotic organisms, of which bacteria are the prime example, no nuclear membrane exists to separate the primary genetic material from the cytoplasm. Indeed, until 1945, scientists did not think that most bacteria had chromosomes, and many biologists doubted whether they had genes, at least in the conventional sense of the term. Such chromosomal material as bacteria do have consists of thin threads that don't look at all like the "colored bodies" cytogeneticists had been seeing under the microscope. These threads do not undergo the rounds of mitosis and meiosis characteristic of chromosomes in higher organisms, and hence bacteria had not been considered as suitable (or even possible) objects for cytological study. Furthermore, it had been known since the end of the nineteenth century that the main chemical constituent of the nucleus is nucleoprotein—a combination of nucleic acid and protein. Even though biologists did not know what the chemical basis of genes was, the structural simplicity of nucleic acids made them appear an unlikely choice. Most geneticists believed that genetic information resided in the protein component of chromosomal material. Bacterial chromosomes, on the other hand, contain little, if any, protein. How, then, could they carry genetic information? Finally, bacteria, reproducing by asexual division, might not even need a genetic apparatus. Could not the entire cell simply replicate itself part by part? All in all, bacteria seemed a poor choice of model to represent the complexity of sexually reproducing, differentiating higher organisms.

The change Demerec observed in 1951 reflected three different but interrelated developments that had been under way throughout the 1940s: it reflected a thoroughgoing transformation in the way genetics was done, in who did genetics, and, finally, in what came to be understood as a gene. In a few years, James Watson would be able to say: "The gene was no longer a mysterious entity whose behavior could be investigated only by breeding experiments. Instead it quickly became a real molecular object about which chemists could think objectively in the same manner as smaller molecules."[11]

Above all, the principal characters of the new drama came from disciplines far removed from classical genetics. They came from biochemistry, from microbiology, from X-ray crystallography, and, perhaps especially, from physics. Few of them had been trained in genetics, and none in cytology. The cytological investigations of the 1930s and 1940s had been enormously influential in convincing biologists of the concrete, physical nature of the gene. But when attention shifted to the gene's molecular character, it did so by leaping over the concerns of cytologists, and to some extent even of the older geneticists.

A leading protagonist of this new way of thinking about genetics was Max Delbrück, a theoretical physicist who had been trained by Niels Bohr. Delbrück is generally regarded as one of the founders of molecular biology—more because of the influence he exerted on the thinking and methodology of the new biologists than for particular discoveries that bear his name. He was steeped in a tradition that seeks understanding in simplicity rather than complexity—that proceeds by isolating and investigating phenomena in their simplest form and treats the variety and plenitude of nature as a distraction, something to be cut through or cleared away in pursuit of general laws. As a physicist, he sought the simplest possible organism available for analysis. And in biology, simplest is likely to mean smallest. That this might mean bypassing some of the very complexities geneticists had been trying to explain was not a problem; it was a virtue.

When Delbrück first came to the United States from Germany in 1937, he went directly to California Institute of Technology, "intent on discovering how his background in physical sciences could be productively applied to biological problems."[12] Work was then already in progress on an organism even smaller than the bacterium—so small, in fact, that it could not even be seen under the microscope. The life history of the bacterial virus had first been described by Felix d'Hérelle in 1926. After attaching itself to a susceptible bacterium, it enters the bacterium where it reproduces itself, and finally the host cell bursts open (lyses), releasing the progeny of the original virus. Since bacterial viruses (or bacteriophages) are too primitive to reproduce themselves without the aid of a host organism, biologists debated whether they were in fact bona fide organisms. Indeed, this marginal status—halfway between chemical molecule and living organism—was what excited Delbrück's interest in the first place. He wrote: "Certain large protein molecules (viruses) possess the property of multiplying within living organisms, [a process] at once so foreign to chemistry and so fundamental to biology."[13] Here surely was the simplest possible system in which to study what was truly essential to the reproductive process. The troublesome question of whether the bacteriophage (or even the bacterium) possessed "genes" was passed over; it was enough that it reproduced.

Eight years later, a somewhat older and slightly chastened Delbrück recalled his youthful enthusiasm in a vein at once amusing and revealing:

> Suppose now that our imaginary physicist, the student of Niels Bohr, is shown an experiment in which a virus particle enters a bacterial cell and 20 minutes later the bacterial cell is lysed and 100 particles are liberated. He will say: "How come one particle has become 100 particles of the same kind in 20 minutes?" That is very interesting. Let us find out how it happens! How does the particle get in to the bacterium? How does it multiply? Does it multiply like a bacterium, growing and dividing, or does it multiply by an entirely different mechanism? Does it have to be inside the bacterium to do this multiplying, or can we squash the bacterium and have the multiplication go on as before? Is this multiplying a trick of organic chemistry which the organic chemists have not yet discovered? Let us find out. This is so simple a phenomenon that the answers cannot be hard to find. In a few months we will know. All we have to do is to study how conditions will influence the multiplication. We will do a few experiments at different temperatures, in different media, with different viruses, and we will know. Perhaps we may have to break into the bacteria at intermediate stages between infection and lysis. Anyhow, the experiments can only take a few hours each, so the whole problem cannot take long to solve.[14]

Delbrück did not solve the "problem of life," but his influence was unmistakable. In the same lecture, his "imaginary physicist" admitted:

> "Well, I made a slight mistake. I could not do it in a few months. Perhaps it will take a few decades, and perhaps it will take the help of a few dozen people. But listen to what I have found. Perhaps you will be interested to join me."[15]

A number of people did listen, in fact had already listened, and were interested in joining with him.

The first of these was the medically trained microbiologist Salvador Luria, another recent arrival from the ravages of German-Italian fascism. Luria had become interested in phage research while still in Italy, and the two men made an easy and natural alliance. To a biological scientist like Luria, it was clear that a central problem of microbiology was to determine whether bacteria were like higher organisms in their genetic apparatus, and especially in their capacity to mutate. It was known that bacteria adapted to different environments, but at the time, many biologists thought that this adaptation was environmentally induced. Luria referred to bacteria as "the last stronghold of Lamarckism."[16] In 1943 he thought of an experiment that could show whether bacterial adaptation was environmentally induced or a product of natural selection operating on spontaneous mutations. He wrote to Delbrück suggesting that they collaborate, and in a matter of months the experiment was done and the results published. Their paper was widely regarded as direct confirmation of the theory of natural selection and, at the same time, as demonstrating that bacteria underwent mutations in the same sense as did higher organisms. Geneticists could now, without reservation, turn their attention to this simple creature that obligingly reproduced itself every twenty minutes. Bacterial genetics had become a legitimate field of study.

A second recruit was an American-born microbiologist named Alfred Hershey. Together, these three constituted the core of what has since become known as the American phage group. In 1969, they shared the Nobel Prize for their contributions to physiology and medicine. Gunther Stent, a molecular biologist turned historian, describes this era as the "romantic phase" of molecular biology—a phase "whose spiritual hallmark was the quest for the physical basis of the gene."[17] To Stent and others, the romantic phase was dominated by Delbrück, Luria, and Hershey.

Growth of the phage group was slow at first, but after Delbrück organized the first annual summer bacteriophage course at Cold Spring Harbor, where the three men had already taken

Gene Conference, Shelter Island, New York, May–June 1949. Standing, left to right: Joshua Lederberg, Francis Ryan, Alfred Sturtevant, Sterling Emerson, Tracy Sonneborn, Milislav Demerec, Alfred Mirsky, Norman Giles, Jack Schultz, William McElroy, Salvador Luria, John Preer, Alfred Hershey. Seated, left to right: Carl Swanson, John Laughnan, Barbara McClintock, Berwind Kaufmann, Lewis Stadler, Curt Stern, Marcus Rhoades. (Courtesy of Marjorie M. Bhavnani.)

to meeting in the summers, it took off exponentially. Delbrück's course put Cold Spring Harbor on the map for an entirely new generation of biologists. As Stent says, its purpose was "frankly missionary: to spread the new gospel among physicists and chemists."[18] Leo Szilard, who some would say exerted at least as important an influence as Delbrück on the recruitment of physicists and on the introduction of the physicist's way of thinking to biology, took the course in 1947. Aaron Novick, an organic chemist who had gone to work with Szilard at the University of Chicago, took it the same summer. Novick called it

> a biology that had been made comfortable for people with backgrounds in the physical sciences. In that three-week course we were given a set of clear definitions, a set of experimental techniques and the spirit of trying to clarify and understand. It seemed to us that Delbrück had created, almost single-handedly, an area in which we could work. . . .[19]

James Watson, then a young first-year graduate student of Luria's, took the phage course in 1948. Delbrück was already an important name to Watson, and he recalls his early meetings with Delbrück with considerable feeling. In those days, there was much talk of the hope shared by Niels Bohr, Erwin Schrödinger, and to some extent Delbrück, that biology would lead to the discovery of "other laws of physics."[20] The frequent reference to quantum mechanics made Watson feel inadequate. "But in the presence of Delbrück I hoped I might someday participate just a little in some great revelation."[21] It turned out that one didn't need to know quantum mechanics to participate in the new way of thinking about genetics; much more important was to learn a certain style that was at once playful, serious, and bold. It meant learning to ask sharp questions, pruning down the luxuriant variety of biology to a few elementary cases that might show the way to a simple explanatory model.

During the weeks at Cold Spring Harbor, these men daily rubbed shoulders with Barbara McClintock. They shared the same grounds, the same dining room, the same lecture halls. But they did not share the same assumptions. Delbrück and Luria held McClintock in the highest regard for her work in classical genetics, but Watson was rapidly learning to look forward rather than back; for him she belonged to a tradition from which he had little to learn. In his reminiscences, "Growing Up in the Phage Group," he mentions McClintock only in passing:

> As the summer passed on I liked Cold Spring Harbor more and more, both for its intrinsic beauty and for the honest ways in which good and bad science got sorted out. On Thursday evenings general lectures were given in Blackford Hall by the summer visitors and generally everyone went, except for Luria who boycotted talks on extra-sensory perception by Richard Roberts and on the correlation of human body shapes with disease and personality by W. Sheldon. On those evenings, as on all others, Ernst Caspari opened and closed the talks, and we marveled at his ability to thank the speakers for their "most interesting presentation."

> Most evenings we would stand in front of Blackford Hall or Hooper House hoping for some excitement, sometimes joking whether we would see Demerec going into an unused room to turn off an unnecessary light. Many times, when it became obvious that nothing unusual would happen, we would go into the village to drink beer at Neptune's Cave. On other evenings, we played baseball next to Barbara McClintock's cornfield, into which the ball all too often went.[22]

If Stent is correct, and the spiritual hallmark of this group was the quest for the physical nature of the gene, it is important to say that that did not necessarily, or even particularly, mean the search for the biochemical identity of the genetic material. Delbrück at least was in search of something more basic, and he thought that biochemistry was not likely to be useful in understanding the really important issues of genetics.[23] As he saw it, biochemists tended to portray the cell as a "sack of enzymes acting on the substrates converting them through various intermediate stages either into cell substance or into waste products";[24] the task they set themselves was to understand how these enzymes are synthesized and how they work. Such an effort, Delbrück thought, is misguided; it seeks to explain "the simple through the complex"[25]—that is, in chemical terms—and, as such, "smacks suspiciously of the program of explaining atoms in terms of complex mechanical models."[26] In other words, instead of seeking the fundamental units of genetics (which Delbrück believed might obey new laws of physics), it mistakenly attempted to work from the top down. The search for the "atoms" of the cell, the proper starting point for explanation in biology, would require far greater attention to the details of cell behavior and a certain theoretical boldness. Delbrück suggested: "It is in this direction that physicists will show the greatest zeal and will create a new intellectual approach to biology which would lend meaning to the ill-used term biophysics."[27]

Fortunately, not everyone shared Delbrück's prejudice. A mere forty miles west of Cold Spring Harbor, investigators at the Rockefeller Institute, who thought rather more highly of the biochemical approach than did Delbrück, were proceeding with another line of attack on the physical (or at least chemical) nature of the gene. Oswald Avery was neither a geneticist nor a physicist. Trained as a physician, he was primarily a microbiologist and a biochemist. He was not in search of new laws of physics or biology, nor even of the physical nature of the gene. Rather, he was intent on finding the substance responsible for the peculiar phenomenon of bacterial transformation—a phenomenon discovered in 1928 by an English physican, Frederick Griffith. Griffith had found that when he mixed two cultures of pneumococcal bacteria, the one virulent but heat-killed, the other live and nonvirulent, the live bacteria somehow acquired the virulent properties of the dead bacteria and transmitted them to their progeny. What was the substance that transmitted the properties of the dead bacteria to the live ones? Only when one understands transformation as an instance of genetic exchange (that is, an exchange of materials that produces a trait henceforth inherited by succeeding generations) does this become a question about genetics. As long as bacteria were not thought to carry genes, identifying the substance responsible for transformation had no obvious implications for the physical nature of the gene. The 1943 experiment of Luria and Delbrück had led some people to think of bacterial genes, but by no means was everyone immediately convinced. Thus, when in 1944 Avery and his colleagues Colin MacLeod and Maclyn McCarty published their analysis showing that the transforming substance was almost certainly DNA, relatively few people leapt to the conclusion that genes were composed of DNA. Even Luria himself was dubious. For one thing, it was hard to imagine that DNA (which Delbrück called a "stupid" molecule because it was composed of only four bases) had enough specificity to carry the enormous quantity of genetic information that living organisms required. Proteins, being so much more complex, and coming in so many varieties, still seemed to many a more obvious choice. Furthermore, nothing in the experiments that Luria and Delbrück were conducting on bacteriophage seemed to implicate DNA in the genetics of that organism. It took eight years and the results of a second crucial experiment to convince the community at large that the genetic material was indeed DNA. In 1952, Al Hershey and Martha Chase at Cold Spring Harbor showed that only the bacteriophage DNA enters the bacterium, while the protein is left outside. After that, things moved very fast. The story of Watson and Crick's discovery in 1954 of the double-helical structure of DNA has been told many times and does not need to be recapitulated here. Suffice it to say that their finding made a colossal impact on biology. It permitted answers, simple beyond anyone's expectation, to a host of questions about the mechanics of inheritance. It immediately answered the question of how genes replicate (each strand of the double helix copies itself by simple chemical bonding of pairs of its basic units, the nucleotide bases), and once the idea was suggested that genetic information was contained in the *sequence* rather than in the *variety* of bases, it was seen that DNA could easily carry the information and specificity required of genes.

Watson and Crick provided the key idea that "the precise sequence of bases is the code that carries the genetic information" in 1953.[28] Its implications were quickly followed up by the physicist George Gamow, who saw the expression of genetic information as a trivial problem in coding. Think of the four possible bases that make up the DNA molecule as letters in one alphabet and the twenty or more possible amino acids that make up a protein molecule as letters in another alphabet. Then, to understand how the DNA "dictates" the construction of particular proteins, one need simply find the code that translates "words" in one alphabet to "words" in the other. Gamow suggested several possible codes, but the three-letter code—in which a sequence of three bases of DNA corresponds to a single aminc acid, and the chain of amino acids that make up the protein molecule corresponds to a long string of base triplets in the DNA—was the one that finally won out.

Most codes allow for two-way translations. But the fundamental hypothesis of molecular biology was that in this case translation could proceed in only one direction. In 1957, Francis Crick dubbed this hypothesis the "central dogma":

> This states that once "information" has passed into protein *it cannot get out again.* [Italics in original.] In more detail, the transfer of information from nucleic acid to nucleic acid, or from nucleic acid to protein may be possible, but transfer from protein to protein, or from protein to nucleic acid is impossible.[29]

From dogma to proof seemed a short step. A few years later Jacques Monod felt entitled to say:

> So what molecular biology has done, you see, is to prove beyond any doubt but in a totally new way the complete independence of the genetic information from events occurring outside or even inside the cell—to prove by the very structure of the genetic code and the way it is transcribed that no information from outside, of any kind, can ever penetrate the inheritable genetic message.[30]

And if genetic information is completely independent of events occurring beyond the genome, no possibility exists for genetic change to be directed by the cell's environment. Once more, the ghost of Lamarck was laid to rest.

Virtually no one questions that the discovery of the structure and function of DNA constituted one of the major revolutions of scientific history. It marked the birth of a new field (molecular genetics), in which the methods and concerns of the older genetics came to seem largely irrelevant. DNA provided a literal as well as figurative scaffolding on which all genetic processes were to be hung. In the minds of molecular geneticists, the earlier categories of genotype and phenotype gave way to DNA and proteins, and the focus of research accordingly shifted from cytology and breeding experiments to biochemistry and the construction of molecular models. The concept of the genes as "beads on a string" was no longer problematic; it was simply unnecessary.

As people gradually gave up thinking of the gene as a "bead" and began to think of it instead as a linear sequence of nucleotide bases, the identity assumed by classical theory of the units of recombination, mutation, and function simply dissolved. The critical experimental work that carried this shift to completion was provided by Seymour Benzer, another physicist initiated into biology by the Cold Spring Harbor phage course. Benzer had accumulated a vast number of bacteriophage mutants that exhibited the same phenotye. By mapping these mutants, he was able to show that even though they were functionally equivalent, recombination could nevertheless occur between many of them: they did not all occur at precisely the same locus. He also showed that the mutations could be formally arranged in a linear array in terms of their frequency of recombination. In 1957, he introduced the term *cistron* to denote the shortest length of genetic material that made up a functional unit (a unit corresponding to a phenotypic trait), and two other terms, *recon* and *muton,* to denote, respectively, the smallest element in the one-dimensional array that can be exchanged in genetic recombination, and the smallest unit of mutation. The cistron was naturally interpreted as a stretch of DNA, and the muton and recon as individual nucleotide bases (or base pairs).

As a result of Benzer's elegant analysis, geneticists felt that the major problems with the classical theory had been resolved. Genes understood as cistrons were no longer indivisible; both mutations and recombination could occur at any point along the DNA, and intragenic recombination ceased to present any conceptual difficulty. For a time, the notion persisted that genes (or cistrons) were separated by extragenic material, but gradually the conviction grew that the chromosome is nothing more than a long continuous chain of DNA, stretches of which constitute separate genes. If so, all mutations must be intragenic.

True, Benzer's analysis was carried out on bacteriophage, but by the late 1950s, bacteriophage and bacteria had become model systems for all of genetics. Monod's assumption that "what is true for *E. coli* is true for the elephant" was thought to apply to bacteriophage as well. Higher organisms like *Dro-*

sophila and maize might be different in some respects, but the hope that the basic mechanisms of genetics were universal carried the day.

Of course, not *all* of the problems of classical genetics had been resolved; as with every scientific revolution, some loose ends and some major conceptual gaps remained. The principal value Goldschmidt had seen in McClintock's work lay in its demonstration that genetic information was not strictly contained in the autonomous gene. The gene as a "bead on a string" could not account for the proposition that the function of a gene might vary with position. The new theory, in which genetic information was contained in the sequence of nucleotide bases, could not better account for such positional dependency than the old theory. If a cistron is a word distinguished by a particular sequence of "letters" (the bases), it must be the same word whatever its place on the chromosome. In that sense, it, too, was a "local" theory. McClintock's work on transposition required the admission of nonlocal, or global, effects. Genetic elements not only changed position, but in each new position, a new function was expressed. In the late 1950s, no one could see a way of accounting for such a phenomenon in terms of DNA sequences.

The question of how such phenomena might be developmentally regulated posed an even greater dilemma. Indeed, for all its successes, molecular genetics was no closer to bridging the gap between genetics and development than the older theory had been. But the enormous success and excitement of the new theory brought with it the inevitable concomitants of hubris and even intolerance. With so many problems being so dramatically solved, who would want to attend to the problems that could not be solved—problems arising in the context of a biology that seemed more and more remote?

CHAPTER 11

Transposition Rediscovered

Molecular biology rescued the gene from decades of ambiguity and confusion. No longer either a hypothetical entity invented to help geneticists order the results of their experiments or a "bead" on the chromosome that cytologists thought they could see through the microscope, the gene was now a distinct chemical entity with a structure suggesting a simple and elegant picture of the mechanics of inheritance. One of the most fundamental questions of genetics had been: How do genes make exact copies of themselves? To this question the double helix provided an instant solution. Watson and Crick proposed that the DNA consists of two intertwined strands that are complementary to each other and joined by a chemical bonding between each pair of complementary bases. That is, each strand specifies the other, much as a photograph specifies a negative. They wrote: "Each chain then acts as a template for the formation onto itself of a new companion chain, so that eventually we shall have two parts of chains, where we only had one before."[1] But the gene must do more than copy itself. If it is to be a true "master molecule" of life, it must oversee the entire construction of the cell; it must transfer the information it carries into the phenotype of the organism. How is it to do this?

By the end of the 1950s, the basic framework of molecular biology was clearly delineated. In addition to making copies of itself, DNA also makes RNA (a nucleic acid structurally similar to DNA) by essentially the same kind of molecular bonding as in replication. Three kinds of RNA are made—all of which mediate between the DNA and proteins. Only one of them (messenger RNA) contains the information coding for the sequence of amino acids making up the actual proteins; the other two (transfer and ribosomal RNA) are structural facilitators for the physical process of protein synthesis. In the succinct language of *Life* magazine, "DNA makes RNA, RNA makes protein, and protein makes us."[2]

Only one serious problem remained. How do cells deriving from a single fertilized egg come to be so different from one another? Presumably they all contain the same DNA. What, then, is the source of the extraordinary diversity, not only in shape but in actual function, that leads one cell to specialize in the production of contractile proteins and another of digestive enzymes? Clearly, in any given cell, only some genes are expressed, even though all are present. What, then, is the mechanism that turns one gene on and another gene off?

• • •

Throughout the 1950s, Cold Spring Harbor was an especially popular meeting place for molecular biologists, and Barbara McClintock had ample opportunity to hear the new results coming in and to share in the general excitement. She listened and she observed, but she maintained a certain critical distance. The gene may no longer have been just a "symbol," but for her the question of the relation of the DNA to the rest of the cell remained crucially at issue. DNA was important, but it was not everything; and the central dogma, in assigning complete autonomy to the DNA, did not adequately account for the regulatory processes underlying differentiation.

McClintock's own work was teaching her, over and over

again, that the genetic apparatus is more labile and flexible than the central dogma allowed. In addition to the *Ds-Ac* system, she had detailed an entirely new system of regulation and control, even more subtle and complex in its organization. She called it the suppressor-mutator system (*Spm*). As before, two controlling elements lay at the source of the observed genetic variation. The first controlling element, in interaction with the second, is capable of effecting a suppression of the gene function (for example, pigmentation) or alternatively of inducing the excision of the second controlling element. In the latter case, gene function (here, the pigmentation) is restored. The two functions of the first controlling element (suppressor and mutator) can undergo independent mutations, indicating that they are coded by separate genes. Furthermore, the mutator not only mediates excision of the second controlling element, but can induce inheritable alterations in its "state." Different "states" express themselves in different levels of overall pigmentation, whereas excision expresses itself in the appearance of dots set off from their background by a distinctly different (usually full) pigmentation. As with the *Ds-Ac* system, these controlling elements could be found not at one standard position on the chromosome but at several positions. Once again, the fact of their "transposability" had been essential to their discovery. McClintock presented the principal features of this system at the Brookhaven Symposium in 1955 and more extensively at the Cold Spring Harbor Symposium in 1956. There she concluded by remarking:

> Controlling elements appear to reflect the presence in the nucleus of highly integrated systems operating to control gene action. The modes of operation of the known two-element systems bring into sharp relief one level of this integration. Other levels are now under investigation. . . . Recognition of the . . . two-element system, then, represents only recognition of the lowest integrative level of those elements in the chromosome complement that are directly concerned with modification of the genome as a whole.
>
> . . . Transposability, which made possible the recognition of con-

> trolling elements in the chromosome complement of maize, may not serve in all cases as a reliable criterion for discrimination between [gene and controlling] elements, because the frequency of its occurrence may be so low, under certain conditions, that detection may be difficult. Nevertheless, . . . [it] would be surprising indeed if controlling elements were not found in other organisms, for their prevalence in maize is now well established.[3]

The existence of a mechanism regulating the rate of production of particular proteins was evident not only to McClintock. On some level, it was evident to everyone, especially to those biochemists who studied the enzymatic adaptation of living cells to their chemical environment. Even *E. coli* appeared capable of adapting its biochemical output in response to the presence or absence of particular chemical substrates in its growth medium. Indeed, so striking was the biochemical adaptability of bacteria to their environment that the phenomenon gave great encouragement to forces that continued—even as late as the 1940s and 1950s—to be hostile to Morgan-Mendelian genetics.

One of the characteristics of scientific development that most plagues historians is the enormous diversity of viewpoints that can continue to persist long after it appears that a consensus has been reached. The difficulty arises not only because consensus is never total, but also because of the fact that consensus always means the consensus of a particular community. Scientists make up many communities, and these communities vary by subject, by methodology, by place, and by degree of influence. Science itself is a polyphonic chorus. The voices in that chorus are never equal, but what one hears as a dominant motif depends very much on where one stands. At times, some motifs appear dominant from any standpoint. But there are always corners from which one can hear minor motifs continuing to sound.

By the middle of the twentieth century, very little opposition to the fundamental arguments of Morgan-Mendelian genetics remained in America, but in Russia, a major attack flourished under the leadership of Lysenko. To the extent that "bourgeois" genetics supported the Darwinian theory of evolution, adaptive (or Lamarckian) evolution required a different genetics. The success of molecular biology confirmed the neo-Darwinian synthesis, but, in the 1950s, molecular genetics was not as influential in continental Europe as it was in England and the United States. Thus while it seemed evident to most American molecular biologists that Luria and Delbrück had defeated the final outpost of Lamarckism with their demonstration of spontaneous mutation in bacteria, biochemists remained preoccupied with the phenomenon of adaptation. Especially in France, this phenomenon appeared to provide support for Lysenko's resurrection of Lamarckism.

Feelings ran high in post–World War II Paris, with science and politics in a volatile mix. American scientists who pride themselves on conscious disengagement from politics find it hard to understand the extent to which French intellectual life tends to be politicized; this was especially true then. In the center of the fray stood Jacques Monod, a hero of the French underground who broke with the Communist Party in 1945, a quintessentially French intellectual who said of himself, "I have to stick to a linear, logical thread—otherwise I am lost,"[4] and a biochemical geneticist who had strong ties with the community of American molecular biologists. Monod was a fierce defendant of the autonomy and the logical self-sufficiency of science. He took upon himself the task of ridding biochemistry of its teleological language; it became his personal crusade to save biology from the corrupting influences of Lysenkoism. As a first step, he proposed substituting the word "induction" for "adaptation." From there, throughout the 1950s, he devoted himself to bringing the problem of biochemical regulation into the framework of molecular biology.

By 1960, he had succeeded. In the fall of that year, together with François Jacob, he published the first account of a molecular mechanism for regulation in *Comptes Rendus.* The following year, a more complete version entitled "Genetic Regulatory Mechanisms in the Synthesis of Proteins" appeared in English.

In this model Jacob and Monod proposed that protein synthesis is regulated not by the structural gene itself (the gene that codes for the protein) but by two other genes—an operator gene lying adjacent to the structural gene, and a regulator gene elsewhere on the chromosome. The regulator gene codes for a repressor (which they suggested was RNA but which later turned out to be protein); the repressor in turn combines with the operator gene to block normal transcription of the structural gene. But in the presence of a particular chemical substrate in the cell that can bind with the repressor, the repressor is neutralized and is no longer capable of combining with the operator. Transcription of the structural gene then becomes reactivated. They called the entire system—structural, regulator, and operator gene—an operon. By way of concluding their 1961 review, they wrote:

> The fundamental problem of chemical physiology and of embryology is to understand why tissue cells do not all express, all the time, all the potentialities inherent in their genome. . . . The discovery of regulator and operator genes, and of repressive regulation of the activity of structural genes, reveals that the genome contains not only a series of blue-prints, but a coordinated program of protein synthesis and the means of controlling its execution.[5]

The consequence of this work was to modify the central dogma in such a way as to vastly expand its scope. It added a crucial form of feedback to the unidirectional command that went from DNA to RNA to protein. The basic feature of the central dogma, however, remained intact; that is, the assumption that "once information got into the protein it could not get out again" still stood. But now, proteins, or other chemicals, could influence at least the rate of flow of information and accordingly regulate the functioning of the entire system. With this proposal, Monod and Jacob provided a satisfying molecular model for how a gene functions in a cell and, as a result, brought the entire domain of biochemical genetics within the scope of molecular biology.

When Barbara McClintock saw Monod's and Jacob's first paper in *Comptes Rendus,* she was overjoyed. Here was an elegant analysis of a bacterial system that bore many similarities to the systems that she had herself worked with in maize. Common to both was the identification of two controlling elements: one adjacent to the structural gene and in direct control of its action, and the other independently situated and, through its effect on the first controlling element, exerting indirect control on the gene. The fact that the controlling elements in her own systems were transposable may have been essential to their initial discovery, but it was not essential to the *operation* of such a control system. Indeed, in one class of gene loci under control of McClintock's *Spm* system, both the element analagous to what Monod and Jacob called the "operator" gene and the *Spm* element itself remain fixed—just as in the bacterial system.

Finally, after a decade of total frustration in her efforts at sharing her knowledge with the world, an echo had rebounded. The similarities between her own ideas about control and regulation and the work of Monod and Jacob were so striking that the latter seemed to provide just the kind of independent confirmation needed to weaken the resistance she had thus far encountered. Now she ought to be able to tell the rest of the story and have it be heard.

Promptly she sent off a paper to *American Naturalist* entitled "Some Parallels Between Gene Control Systems in Maize and in Bacteria," in which, after outlining the basic features of the *Spm* system, she concluded:

> It is expected that such a basic mechanism of control of gene action will be operative in all organisms. In higher organisms, lack of means of identifying the components of a control system of this type may be responsible for delay in recognition of their general prevalence, even though there is much genetic and cytological evidence to indicate that control systems do exist. It is anticipated, however, that control systems exhibiting more complex levels of integration will be found in the higher organisms.[6]

She also gave a seminar at Cold Spring Harbor calling attention to the same parallels. A number of people in the audience were quite excited, but it was the work of Monod and Jacob, and the possibilities that work suggested, that excited them. The receptivity to her own work, which she had thought would follow, did not come. Controlling elements in maize remained incomprehensible.

Part of the problem was that the idea of transposition remained simply indigestible. But by now, she was genuinely puzzled. Why was transposition so heretical? By this time, biologists had become amply familiar with the fact that viral DNA was readily inserted into (and released from) bacterial DNA, often carrying pieces of the bacterial chromosome with it. In some cases, more than one site of insertion had been identified. What, then, was so different about the release and insertion of pieces of resident DNA? It seemed to her simply illogical to consider one reasonable and the other not. What kind of invisible barrier kept transposition beyond the range of acceptability?

It was not the first time she had encountered such a seeming lack of logic in her colleagues. The cause? For McClintock it was what she calls "tacit assumptions"—an implicit adherence to models that prevents people from looking at data with a fresh mind. These tacit assumptions impose unconscious boundaries between what is thinkable and what is not. Even glaring lapses of logic become invisible: "They didn't know they were bound to a model, and you couldn't show them . . . even if you made an effort." She found no difficulty with young scientists, but she felt that with too many of her colleagues age brought on a hardening of the mental arteries. Years of reading the literature, of listening to seminars, make it more rather than less difficult for them to become aware of their hidden assumptions, to hear something new. The unfamiliar becomes increasingly unthinkable, and people forget that their interpretations of data are subject to change; they forget that theories and models come and go. The greatest danger, she finds, comes when people try to explain everything on the basis of what they think they know. "That's why models, when they first begin to be promulgated, are so bothersome to me." Out of sheer enthusiasm, the model gets mistaken for reality. "The central dogma," she says, "is one of these."

The operon theory of Monod and Jacob was a long way from explaining all forms of regulation. What might work for the bacterium would not necessarily work for higher organisms. McClintock points out: "The eukaryotes are made up of a lot of cells, and no two cells in different parts of the organism can be doing the same thing. Therefore, there must be controls that are very different from what you get in the bacteria. Bacteria are highly evolved organisms. Their operons are just superb—extraordinary economy. [But] we don't use that kind of economy in the higher organisms."

Molecular biologists weren't thinking about eukaryotes. "They had no feel for what these cells had to undergo in development," according to McClintock. "Organisms can do all types of things; they do fantastic things. They do everything that we do, and they do it better, more efficiently, more marvelously. . . . Trying to make everything fit into set dogma won't work. . . . There's no such thing as a central dogma into which everything will fit. It turns out that any mechanism you can think of, you will find—even if it's the most bizarre kind of thinking. Anything . . . even if it doesn't make much sense, it'll be there. . . . So if the material tells you, 'It may be this,' allow that. Don't turn it aside and call it an exception, an aberration, a contaminant. . . . That's what's happened all the way along the line with so many good clues."

The challenge for investigators in every field is to break free of the hidden constraints of their tacit assumptions, so that they can allow the results of their experiments to speak for themselves. "I feel that much of the work is done because one wants to impose an answer on it," McClintock says. "They have the answer ready, and they [know what they] want the material to tell them." Anything else it tells them, "they don't really recognize as there, or they think it's a mistake and throw it out. . . . *If you'd only just let the material tell you.*"

Molecular biologists were in even less of a mood to listen to McClintock than they were to "listen to the material." The message they got from their own experiments was that their models, and model building, were overwhelmingly successful. She may have felt that she was in closer touch with biological reality, but from the point of view of many of her colleagues at Cold Spring Harbor, she seemed increasingly out of touch. Wedded to old-fashioned concepts, personally eccentric, her insights seemed less and less to belong to what they thought of as science. She seemed to delight in espousing or at least entertaining heretical positions. Privately, she maintained a serious interest in Buddhist thought, and, as Adrian Srb recently commented, "if other scientists reject the idea that there are UFOs, McClintock probably would not decide until she could either prove or disprove their existence."[7]

The seminar McClintock gave in 1960 was her last attempt to explain her work to her colleagues at Cold Spring Harbor. She felt she had tried hard enough. Time passed, and her research continued—as ever, a reliable source of consolation and sustenance. She would be rewarded by her understanding, if not by her colleagues.

As she had long since discovered, if one looked hard enough and carefully enough, a single organism would reveal its secrets. It would tell you not of one but of many mechanisms it had evolved to regulate the expression of genes—mechanisms that enabled its cells to produce exactly what was needed, when it was needed. It was an exquisitely balanced timepiece that seemed to be capable of whatever readjustments circumstances required. Some mechanisms involved massive reorganization of the genome; others merely modulated the expression of genes without changing the DNA composition. Though she couldn't provide molecular explanations for any of these events, she could plainly see their effects. There was no question that they occurred. The problem was to persuade others, and this she could not do.

If she had felt isolated in her views before, now her isolation took on new scope. The success of molecular biology brought a vision of unprecedented order into biology, and in this vision, which placed so high a priority on the power of simple models to account for the complexity of living phenomena, there was little room for phenomena that defied such accountability. Molecular biologists were a new breed of life scientists. Young, irreverent, and overwhelmingly confident, they were turning biology into what they regarded as a bona fide science—a science more like physics than even the most ardent experimentalists had ever thought possible.

Up until the end of the nineteenth century, biology had been primarily an observational science; biologists had sought to capture the mysteries of nature by documentation and description, rather than by a priori explanation. The early twentieth century saw the transformation of biology into an experimental science. But for many researchers, commitment to the integrity of the organism, and a reverence for the opulent variety of nature, remained. Not until the advent of molecular biology did the final break with earlier tradition occur. The long-standing tension between the organism as a whole and its constituent physico-chemical parts appeared at last to be relieved. Biology could now be seen as a science of molecular mechanics, rather than of living organisms, or even of "living machines." Old-timers like McClintock who were still committed to the inherent complexity and mystery of life were expected to step aside. A new generation was creating a new biology.

McClintock was now definitely outnumbered at Cold Spring Harbor, and the only options she knew were retreat and further withdrawal. Fortunately, in the late 1950s, a new option materialized. It came in the form of an invitation from the National Academy of Sciences—an invitation that, in more ways than one, took her far afield from her research on regulation and control. For our purposes, the story of that detour can be told very briefly.

The National Academy of Sciences had identified a serious threat to the population of indigenous maize in Central and South America. With the rapid spread of agricultural corn, the indigenous strains would soon be lost if they were not collected and preserved. A committee was set up, and McClintock was asked if she could help train local cytologists needed to carry out the project. She promptly agreed—mainly out of a sense of obligation, but perhaps, too, from a desire to get away. It meant a vacation, travel (she quickly learned to speak a creditable Spanish), and, as it turned out, a chance to think creatively about a whole new set of issues. What she soon saw was that by studying the geographical distribution of particular chromosomal types, once again she began to discern patterns. And she recognized that from these patterns, it would be possible to trace the patterns by which people settled and traded in the Americas; that is, a reconstruction of the biological history of the maize plant would permit a reconstruction of the migratory history of humans. The crucial point is that, unlike other grains, corn grows only where humans live. Because the corn seeds are tightly enclosed in the husk, the plant's propagation is entirely dependent on human intervention. Therefore, when McClintock saw that the variations in chromosomal constitution fell into a geographic pattern, reflecting successive rounds of hybridization, she knew that the data would be of enormous interest to anthropologists.

After spending two winters in Central and South America, between 1958 and 1960, she left the rest of the data for her colleagues to collect. Her own final report of what amounted to over a decade-long study did not appear until 1978. In the meantime, she continued her work on transposition with dogged determination. "I knew I was right," she says. In 1965, she made a fourth attempt to describe her findings, at the Brookhaven Symposium, but with little effect. New honors from the larger world of biology accorded her during this period—Cornell appointed her an Andrew White Professor-at-Large (a nonresident appointment) in 1965, the National Academy of Sciences selected her for the Kimber Genetics Award in 1967, and in 1970 she received the National Medal of Science—brought her some encouragement, but could not console her over the rejection of her most important discoveries. But finally, in the mid- to late 1970s, when the face of molecular

Barbara McClintock in her laboratory at Cold Spring Harbor, 1963. (Permission of Marjorie M. Bhavnani.)

biology had grown vastly more complex, the patterns she had seen in her corn kernels slowly began to become visible to others.

The miracle of life is that, despite the best grip we can get on reality, it continuously manages to surprise us. The beauty of science is that, notwithstanding all our tacit assumptions, these surprises can get through. Beginning in the mid-1960s—slowly at first but, at last, incontrovertibly—a number of new experimental findings, in the very organisms that molecular biologists were studying, began to undermine their confidence in the stability of the genome. To be sure, organisms reproduce themselves with remarkable fidelity. But mounting evidence forced recognition of a wide range of circumstances under which the genome undergoes rearrangement.

Given the prevailing confidence in the central dogma, the fact that the breakthrough came from molecular biology itself may seem somewhat surprising. But at the same time it is difficult to imagine its having been otherwise. In the transition from classical to molecular genetics, the very definition of what constitutes compelling evidence had changed. Given the degree of confidence molecular biologists had acquired in their own methodology, and the corresponding lack of understanding of the experimental work of their predecessors, it seems inevitable that their tacit assumptions could be effectively challenged only on their own turf.

The first sign of something new was not by itself seen as radical, but, by undermining the distinction between transduction and transposition, it did help to prepare the way for subsequent challenges. Transduction was a well-known phenomenon in which bacteriophage proved capable of carrying pieces of genetic material from one bacterial chromosome to another. What distinguished it from transposition was that the sites of bacteriophage insertion, and consequently of possible deletion or insertion of bacterial genes, appeared to be uniquely defined, and fixed. But in a little-noticed paper, A. L. Taylor had shown in 1963 that a certain bacteriophage, called mμ, could insert itself into the bacterial chromosome at a very large number of sites, perhaps even at random.[8] This meant that, in moving from one bacterium to another, or even from one site to another on the same chromosome, mμ could serve as an agent of a kind of induced genetic "transposition," although Taylor himself did not call it that. However, the word *was* used a few years later, albeit without explicit reference to McClintock, by Jonathan Beckwith, Ethan Signer, and Wolfgang Epstein in their 1966 report of a similar phenomenon involving the "F factor" (a viruslike particle that replicates autonomously in the bacterial cell).[9]

The next step was more startling. By the late 1960s biologists in several laboratories were hard at work trying to understand a new class of mutations occurring in the operons of *E. coli.* These mutations, which seemed to constitute a large fraction of all spontaneously occurring mutations, are unusual in several respects. First, they not only abolish the function of mutated genes, but exert strong inhibiting (or, in one case at least, promoting) effects on genes downstream from the affected gene. Furthermore, the fact that they are capable of spontaneous reversion but do not respond to known mutagenic agents suggested that they are due to some kind of chromosomal aberration. It was soon found that they are caused by the insertion of one of a small group of specific segments of DNA (insertion sequences) into a structural or regulator gene. These insertion sequences are not foreign DNA, like bacteriophage, but rather material displaced from elsewhere on the bacterial chromosome. Their insertion into a gene signals a mutation; their excision, a reversion. Excision is usually precise, leading to the restoration of normal gene function. But occasionally excision is imprecise, and the insertion elements pick up some of the adjacent genetic material. They can then carry the new material with them to a new position, inserting it in either the same or the reversed orientation. By this mechanism, insertion elements can cause deletions, translocations, and inversions—in short, just the kinds of genetic rearrangement McClintock had identified as a consequence of transposition in maize. The function of insertion elements was (and still is) unclear, but from their first identification, it was clear that they could turn genes on and off. It appeared, therefore, that they might be implicated in the phenomenon of regulation and control.

Soon after the identification of insertion sequences, an even more dramatic instance of genetic mobility was found in the bacterium *Salmonella typhimurium,* this time with direct medical implications. It had been known for some time that the genes responsible for the bacterium's drug resistance were capable of spreading at an alarming rate, and in the mid-1970s, a clue to their rapid dissemination was found in their mobility within the chromosome. Normally residing on an extra-chromosomal fragment of DNA called a plasmid, these genes were first seen to move on the back of a bacteriophage. But what might have seemed to be merely another instance of transduction (similar to that found in mμ) was soon demonstrated to be a phenomenon that could proceed independently of the bacteriophage. The drug-resistant genes, either singly or in groups, seemed to move at will: from the plasmid to a bacteriophage, from a bacteriophage to the bacterial chromosome, from one position on the chromosome to another, from there to another bacteriophage that could then carry it to yet another bacterium. Molecular analysis revealed that the genes coding for drug resistance were embedded in elements with a characteristic structure of a kind that suggested a possible mechanism for their mobility. Bounding the genes on both sides were sequences of DNA that were typically inverted repeats of each other (or sometimes direct repeats) that could bind to each other by homologous base-pairing, thereby forming a characteristic stem and loop (or "lollipop") structure that could actually be seen under the electron microscope. Indeed, it was the observation of these structures that constituted the first evidence of repeating sequences at the two ends. Eventually, the whole element came to be called a transposon.

Insertion elements, drug-resistant genes, and the bacteriophage mμ all had in common the fundamental feature that they could insert themselves into the bacterial chromosome in the absence of normal recombination, and they could do so with little or no specificity. In all three cases, genetic rearrangement resulted. Investigation of the molecular structure of these elements soon revealed even stronger commonalities. The sequences of DNA bounding the drug-resistant genes were found to be similar and, in some cases, identical to those bounding the insertion sequences. Accordingly, it seemed reasonable to suppose that insertion sequences, residing in multiple copies on the bacterial chromosomes, might serve as sites for the integration of genes bounded on both sides by homologous stretches of

DNA. It was even suggested that they constitute "joints for the modular construction of chromosomes."[10]

The critical function of the terminal repeats bounding the transposons was further substantiated by the discovery that a pair of bacteriophages mμ could, by attaching themselves at opposite ends to a piece of bacterial DNA, mimic the same structure and the same effect. The evolutionary implications were obvious—perhaps especially so in the case of the drug-resistance genes. A mechanism that permits so much flexibility and ease of dissemination confers an enormous evolutionary advantage. And even though it had seemed hardly conceivable a few years earlier, now it seemed logical that bacteria, given that they are subject to such rapid changes in environmental pressures, might evolve a mechanism that would so greatly increase their adaptability.

By this time, excitement about movable genetic elements (or "jumping genes," as they were sometimes called) approached epidemic proportions. But the documentation of transposition in bacteria by no means constituted a vindication of McClintock's discovery of transposable elements in maize. Ironically, once the phenomenon was firmly established in lower organisms, the question became: Does transposition also occur in higher organisms? Peter Starlinger and Heinz Saedler were the first to draw the analogy between insertion sequences (IS) and McClintock's work (as early as 1972), but the connection did not truly catch on until 1976, when at a Cold Spring Harbor meeting on "DNA Insertion Elements, Plasmids, and Episomes," explicit acknowledgement of McClintock was made in introducing the term "transposable elements" to refer to all "DNA segments which can insert into several sites into a genome."[11] The proceedings of that meeting were published a year later. Even then, however, no one was quite sure. Saedler himself, one of the most enthusiastic advocates of the parallels between bacteria and maize, wrote:

> Whatever the role of IS elements in the evolution of chromosomes and plasmids may be, it is noteworthy that formally analo-

> gous elements are also observed in eukaryotic organisms such as *Zea mays, Drosophila melanogaster,* and others. Whether these eukaryotic elements have any relationship to the known IS elements of prokaryotic origin is open to speculation.[12]

The reasons for this hesitation were several. For one thing, the very nature of the molecular investigations made the analogy difficult to pin down. Transposition in bacteria had been identified by the structural properties of the insertion elements; McClintock's work was based entirely on their functional consequences. Without access to the same kinds of identifying features, it was impossible to have confidence in their similarities. Furthermore, molecular geneticists remained largely ignorant of the intricacies of maize genetics, and the arguments McClintock employed were inaccessible to them. A big step forward was made in 1977 when Patricia Nevers and Heinz Saedler familiarized themselves with the details of McClintock's system and published a carefully reasoned paper proposing a molecular model of eukaryotic controlling elements that exploited all the similarities between the bacterial and maize phenomena and that, accordingly, made maize more accessible to their colleagues.[13]

But what was probably the greatest difference of all between the two systems remained, and that had to do with function. The principal significance of transposition in maize lay in its regulatory function. McClintock had named her transposition elements "controlling elements" because of the role they played in regulating their own function and the function of neighboring genes. She had shown them capable of regulating the precise timing of genetic function—according to a timetable that was in part determined by the number of controlling elements present. Nothing as subtle had been demonstrated in bacterial transposition. Insertion elements might turn genes on and off, but even after it had been shown that their effect could depend on the orientation in which they were inserted, there was little to argue against the proposition that they acted merely by disrupting the normal functioning of the genetic sequence they entered.

Perhaps the closest thing to a "controlling element" found in bacteria was the "flip-flop" switch in *Salmonella.* It had been known for some time that two types of bacteria flagella are alternatively produced by this bacterium; in 1978, the switch responsible for a change from one to the other was identified as a specific sequence of DNA capable of periodically reversing its orientation. In one orientation, the switch was "on" and one type of flagellum was produced; in the other orientation, the switch was "off" and the other type was produced. But even with so dramatic an instance of regulation, no developmental significance could be inferred from the work on bacteria, and most workers remained skeptical that such significance could be attributed to any form of transposition. In short, transposition was regarded as an essentially aberrant phenomenon—one that might have evolutionary consequences but that was not thought of as having implications for developmental organization.

No one was more conscious of the difference in focus than McClintock herself. In part, that difference reflected the continuing disparity between her own interests and those of her colleagues in genetics. She was primarily interested in function and in organization; they were primarily interested in mechanism. But also, it was in part a reflection of the differences between the organisms studied. Bacteria like *E. coli* and *Salmonella* do not have a developmental cycle. Higher organisms do. It is therefore hardly surprising that the developmental consequences of transposition did not begin to emerge until biologists started to look for, and find, similar phenomena in eukaryotic systems.

Although evidence that transposition is implicated in the development of *Drosophila* was reported early on by Melvin Green, yeast was the first higher organism to attract wide attention for exhibiting such an effect. Two different developmental stages, corresponding to sexually complementary functions, were shown to result from the physical insertion of one of two genes, from different places on the genome, into a third locus (the mating locus). In the last couple of years, an even closer parallel between yeast and maize has emerged from studies by

Gerald Fink and his coworkers of unstable mutants in one of the loci responsible for the synthesis of the amino acid histidine.[14] So extensive are the similarities between Fink's system and the system McClintock called *Spm* (for suppressor-mutator) that Fink has adopted the same name for his own system. In maize, the *Spm* system has two components: the first (a "receptor") inserts into or near a gene, causing a mutant phenotype, and the second (a "regulator"), generally remote from the gene in question, controls, or regulates, the activity of the first. The regulator component controls both the degree to which the first component suppresses (or enhances) the function of the gene in or near which it resides (suppressor activity) and also the frequency of excision of the first component (mutator activity). The yeast *Spm* system has the same two components. Functionally, the two systems are of a kind, but Fink is careful to add that a common nomenclature need not imply a common mechanism.

Moving a little further up the evolutionary ladder, researchers have now begun to find an even greater abundance of "jumping genes" in *Drosophila*; some of these appear to be directly implicated in development. In a group of genes called the bithorax complex (so named because it controls the development of the segmented body of the insect), transposition has been identified as the mechanism underlying a number of mutations that dramatically affect the morphology of the fly. A genetic element moving from one site to another can cause a change in developmental instructions leading to the formation of, say, an extra leg instead of a wing, or in other mutants, a wing instead of a piece of the compound eye. Thus far, almost all the developmental phenomena that have been linked to transposition are abnormalities (as was also the case in maize), but some biologists are beginning to speculate that genetic rearrangement may be a feature of normal eukaryotic development as well. Perhaps the best available evidence supporting this view comes from the study of antibody production in mammalian cells. In a number of laboratories, researchers have demonstrated that the generation of diverse populations of antibody molecules depends on the routine occurrence of genetic rearrangement during the course of development. In an article reviewing the recent evidence, James A. Shapiro has written:

> It is almost superfluous to say that our understanding is virtually nil of how cell division is connected to any regulatory event. . . . However, it is now clear that such connections do exist and that the bizarre patterns on corn kernels controlled by mobile genetic elements may be typical of normal development processes rather than exceptional.[15]

Shapiro follows this article by another, one of the major points of which is to emphasize that "McClintock's and other studies in classical cytogenetics must form the essential background for interpreting the flood of data that has resulted from technological advances in analysis of DNA sequences and chromosome fine structure."[16]

Drawing encouragement from such support, McClintock has now become more outspoken in recent years about the implications she sees in transposition. In a paper she presented at the Stadler Symposium in 1978, entitled "Mechanisms That Rapidly Reorganize the Genome," she went beyond the question of developmental control and regulation to discuss the more general occurrence of innate mechanisms for restructuring the genome, mechanisms that are called into action by internal and external stress.[17] She reviews evidence for specific mechanisms they have evolved for responding to traumatic stress that "could provide newly organized genomes with orderly operating gene-control systems while still retaining those components that again can respond to stress."[18] The evolutionary implications are vast, if somewhat obscure. But, as she concludes in a 1980 paper:

> There is little doubt that genomes of some if not all organisms are fragile and that drastic changes may occur at rapid rates. These can lead to new genomic organizations and modified controls of

Barbara McClintock, Cold Spring Harbor, 1980. (Permission of Cold Spring Harbor Laboratory Research Library Archives/Photographer Herb Parsons.)

> type and time of gene expression. . . . Since the types of genome restructuring induced by such elements know few limits, their extensive release, followed by stabilization, could give rise to new species or even new genera.[19]

The importance of McClintock's work on transposition is no longer denied by anyone, but resistance to the more radical dimensions of her vision remains deep; many biologists regard them as pure (if not wild) speculation. So major a challenge to conventional views requires more evidence than they feel yet exists. McClintock's own evidence for the developmental consequences of transposition is, even now, understood by only a very few. And her arguments for innate mechanisms for genetic responses to stress are understood by even fewer. Among others, a disparaging banter continues: to them "McClintockian" has become a code word for unscientific.

Nonetheless, even the skeptics must acknowledge this much: the genome is not a static entity, but a complex structure in a state of dynamic equilibrium. And transposable elements—all having the same structural organization—are a common feature of higher and lower organisms alike. They are neither a dubious nor an isolated phenomenon. As Melvin Green of the University of California at Davis says, "they are everywhere, in bacteria, yeast, *Drosophila,* and plants. Perhaps even in mice and men."[20] But just what they imply for genetic organization, development, and evolution remains a subject of ongoing debate. More conservative thinkers regard them as an interesting, even surprising, new mechanism that must be added to the basic repertoire of cellular mechanisms, but not as one that fundamentally challenges either the autonomy or the primacy of the DNA. True, transposition allows for more rapid evolution than had been previously thought, but to them genetic change remains random, and the central dogma and the theory of natural selection remain essentially intact. To these thinkers, the primary interest of transposition lies in its mechanism; its occurrence does not imply a revolution in biological thought. Others, however—and the number is growing—see a fundamental contradiction between the dynamic properties of the chromosome now emerging and the earlier static view. But no one can yet quite see how to resolve this contradiction. Does it require rethinking the internal relations of the genome, exploring ways in which internal feedback can generate programmatic change? Or does it require rethinking the relation between the genome and its environment, exploring the ways in which the

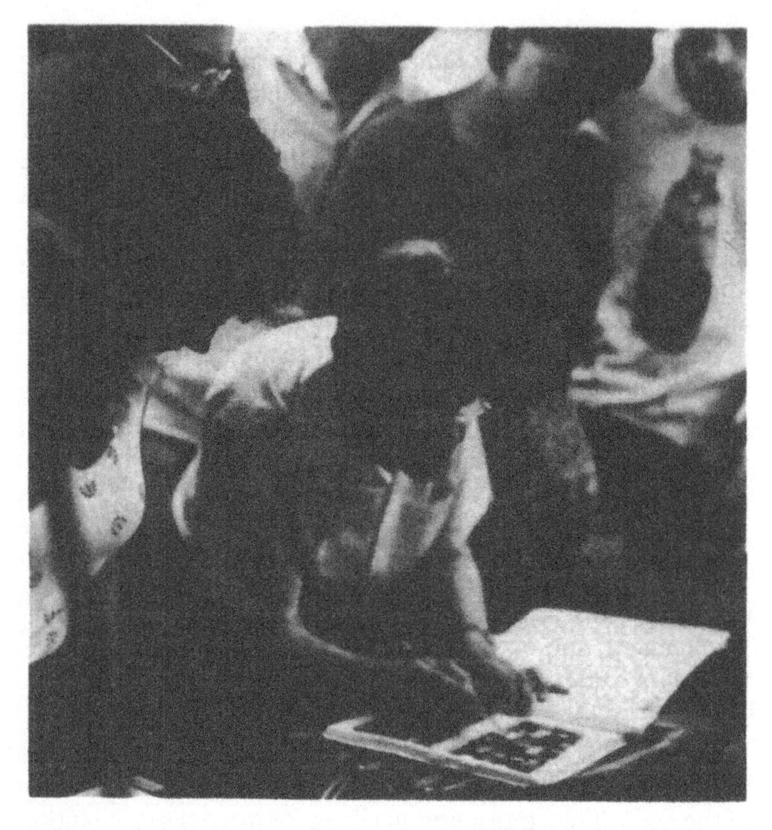

McClintock's Plant Course at Cold Spring Harbor, 1981. (Permission of Cold Spring Harbor Laboratory Research Library Archives/Photographer Herb Parsons.)

DNA can respond to environmental influences? Or does it require both?

Without question, the genetic apparatus is the guarantor of the basic stability of genetic information. But equally without question, it is a more complex system, with more complex forms of feedback, than had been previously thought. Perhaps the future will show that its internal complexity is such as to enable it not only to program the life cycle of the organism, with fidelity to past and future generations, but also to reprogram itself when exposed to sufficient environmental stress—thereby effecting a kind of "learning" from the organism's experience. Such a picture would be radical indeed, and it would be one that would do justice to McClintock's vision: it would imply a concept of genetic variation that is neither random nor purposive—and an understanding of evolution transcending that of both Lamarck and Darwin.

CHAPTER 12

A Feeling for the Organism

There are two equally dangerous extremes—
to shut reason out, and to let nothing else in.

PASCAL

If Barbara McClintock's story illustrates the fallibility of science, it also bears witness to the underlying health of the scientific enterprise. Her eventual vindication demonstrates the capacity of science to overcome its own characteristic kinds of myopia, reminding us that its limitations do not reinforce themselves indefinitely. Their own methodology allows, even obliges, scientists to continually reencounter phenomena even their best theories cannot accommodate. Or—to look at it from the other side—however severely communication between science and nature may be impeded by the preconceptions of a particular time, some channels always remain open; and, through them, nature finds ways of reasserting itself.

But the story of McClintock's contributions to biology has another, less accessible, aspect. What is it in an individual scientist's relation to nature that facilitates the kind of seeing that eventually leads to productive discourse? What enabled McClintock to see further and deeper into the mysteries of genetics than her colleagues?

Her answer is simple. Over and over again, she tells us one must have the time to look, the patience to "hear what the material has to say to you," the openness to "let it come to you." Above all, one must have "a feeling for the organism."

One must understand "how it grows, understand its parts, understand when something is going wrong with it. [An organism] isn't just a piece of plastic, it's something that is constantly being affected by the environment, constantly showing attributes or disabilities in its growth. You have to be aware of all of that. . . . You need to know those plants well enough so that if anything changes, . . . you [can] look at the plant and right away you know what this damage you see is from—something that scraped across it or something that bit it or something that the wind did." You need to have a feeling for every individual plant.

"No two plants are exactly alike. They're all different, and as a consequence, you have to know that difference," she explains. "I start with the seedling, and I don't want to leave it. I don't feel I really know the story if I don't watch the plant all the way along. So I know every plant in the field. I know them intimately, and I find it a great pleasure to know them."

This intimate knowledge, made possible by years of close association with the organism she studies, is a prerequisite for her extraordinary perspicacity. "I have learned so much about the corn plant that when I see things, I can interpret [them] right away." Both literally and figuratively, her "feeling for the organism" has extended her vision. At the same time, it has sustained her through a lifetime of lonely endeavor, unrelieved by the solace of human intimacy or even by the embrace of her profession.

Good science cannot proceed without a deep emotional investment on the part of the scientist. It is that emotional investment that provides the motivating force for the endless hours of intense, often grueling, labor. Einstein wrote: " . . . what deep longing to understand even a faint reflexion of the reason revealed in this world had to be alive in Kepler and Newton so that they could in lonely work for many years disentangle the mechanism of celestial mechanics?"[1] But McClintock's feeling for the organism is not simply a longing to behold the "reason revealed in this world." It is a longing to embrace the world in its very being, through reason and beyond.

For McClintock, reason—at least in the conventional sense of the word—is not by itself adequate to describe the vast complexity—even mystery—of living forms. Organisms have a life and order of their own that scientists can only partially fathom. No models we invent can begin to do full justice to the prodigious capacity of organisms to devise means for guaranteeing their own survival. On the contrary, "anything you can think of you will find." In comparison with the ingenuity of nature, our scientific intelligence seems pallid.

For her, the discovery of transposition was above all a key to the complexity of genetic organization—an indicator of the subtlety with which cytoplasm, membranes, and DNA are integrated into a single structure. It is the overall organization, or orchestration, that enables the organism to meet its needs, whatever they might be, in ways that never cease to surprise us. That capacity for surprise gives McClintock immense pleasure. She recalls, for example, the early post-World War II studies of the effect of radiation on *Drosophila*: "It turned out that the flies that had been under constant radiation were more vigorous than those that were standard. Well, it was hilarious; it was absolutely against everything that had been thought about earlier. I thought it was terribly funny; I was utterly delighted. Our experience with DDT has been similar. It was thought that insects could be readily killed off with the spraying of DDT. But the insects began to thumb their noses at anything you tried to do to them."

Our surprise is a measure of our tendency to underestimate the flexibility of living organisms. The adaptability of plants tends to be especially unappreciated. "Animals can walk around, but plants have to stay still to do the same things, with ingenious mechanisms. . . .Plants are extraordinary. For instance, . . . if you pinch a leaf of a plant you set off electric pulses. You can't touch a plant without setting off an electric pulse. . . . There is no question that plants have [all] kinds of sensitivities.

They do a lot of responding to their environment. They can do almost anything you can think of. But just because they sit there, anybody walking down the road considers them just a plastic area to look at, [as if] they're not really alive."

An attentive observer knows better. At any time, for any plant, one who has sufficient patience and interest can see the myriad signs of life that a casual eye misses: "In the summertime, when you walk down the road, you'll see that the tulip leaves, if it's a little warm, turn themselves around so their backs are toward the sun. You can just see where the sun hits them and where the sun doesn't hit. . . . [Actually], within the restricted areas in which they live, they move around a great deal." These organisms "are fantastically beyond our wildest expectations."

For all of us, it is need and interest above all that induce the growth of our abilities; a motivated observer develops faculties that a casual spectator may never be aware of. Over the years, a special kind of sympathetic understanding grew in McClintock, heightening her powers of discernment, until finally, the objects of her study have become subjects in their own right; they claim from her a kind of attention that most of us experience only in relation to other persons. "Organism" is for her a code word—not simply a plant or animal ("Every component of the organism is as much of an organism as every other part")—but the name of a living form, of object-as-subject. With an uncharacteristic lapse into hyperbole, she adds: "Every time I walk on grass I feel sorry because I know the grass is screaming at me."

A bit of poetic license, perhaps, but McClintock is not a poet; she is a scientist. What marks her as such is her unwavering confidence in the underlying order of living forms, her use of the apparatus of science to gain access to that order, and her commitment to bringing back her insights into the shared language of science—even if doing so might require that language to change. The irregularities or surprises molecular biologists are now uncovering in the organization and behavior of DNA are not indications of a breakdown of order, but only of the inadequacies of our models in the face of the complexity of nature's actual order. Cells, and organisms, have an organization of their own in which nothing is random.

In short, McClintock shares with all other natural scientists the credo that nature is lawful, and the dedication to the task of articulating those laws. And she shares, with at least some, the additional awareness that reason and experiment, generally claimed to be the principal means of this pursuit, do not suffice. To quote Einstein again, "... only intuition, resting on sympathetic understanding, can lead to [these laws]; ... the daily effort comes from no deliberate intention or program, but straight from the heart."[2]

A deep reverence for nature, a capacity for union with that which is to be known—these reflect a different image of science from that of a purely rational enterprise. Yet the two images have coexisted throughout history. We are familiar with the idea that a form of mysticism—a commitment to the unity of experience, the oneness of nature, the fundamental mystery underlying the laws of nature—plays an essential role in the process of scientific discovery. Einstein called it "cosmic religiosity." In turn, the experience of creative insight reinforces these commitments, fostering a sense of the limitations of the scientific method, and an appreciation of other ways of knowing. In all of this, McClintock is no exception. What is exceptional is her forthrightness of expression—the pride she takes in holding, and voicing, attitudes that run counter to our more customary ideas about science. In her mind, what we call the scientific method cannot by itself give us "real understanding." "It gives us relationships which are useful, valid, and technically marvelous; however, they are not the truth." And it is by no means the only way of acquiring knowledge.

That there are valid ways of knowing other than those conventionally espoused by science is a conviction of long standing for McClintock. It derives from a lifetime of experiences that science tells us little about, experiences that she herself could no more set aside than she could discard the anomalous pattern on a single kernel of corn. Perhaps it is this fidelity to her own experience that allows her to be more open than most other scientists about her unconventional beliefs. Correspondingly, she is open to unorthodox views in others, whether she agrees with them or not. She recalls, for example, a lecture given in the late 1940s at Cold Spring Harbor by Dick Roberts, a physicist from the Carnegie Institution of Washington, on the subject of extrasensory perception. Although she herself was out of town at the time, when she heard about the hostile reaction of her colleagues, she was incensed: "If they were as ignorant of the subject as I was, they had no reason for complaining."

For years, she has maintained an interest in ways of learning other than those used in the West, and she made a particular effort to inform herself about the Tibetan Buddhists: "I was so startled by their method of training and by its results that I figured we were limiting ourselves by using what we call the scientific method."

Two kinds of Tibetan expertise interested her especially. One was the way the "running lamas" ran. These men were described as running for hours on end without sign of fatigue. It seemed to her exactly the same kind of effortless floating she had secretly learned as a child.

She was equally impressed by the ability that some Tibetans had developed to regulate body temperature: "We are scientists, and we know nothing basically about controlling our body temperature. [But] the Tibetans learn to live with nothing but a tiny cotton jacket. They're out there cold winters and hot summers, and when they have been through the learning process, they have to take certain tests. One of the tests is to take a wet blanket, put it over them, and dry that blanket in the coldest weather. And they dry it."

How were they able to do these things? What would one need to do to acquire this sort of "knowledge"? She began to look at related phenomena that were closer to home: "Hypnosis also had potentials that were quite extraordinary." She began to believe that not only one's temperature, but one's circulation, and many other bodily processes generally thought to be autonomous, could be brought under the influence of mind. She was convinced that the potential for mental control revealed in hypnosis experiments, and practiced by the Tibetans, was something that could be learned. "You can do it, it can be taught." And she set out to teach herself. Long before the word "biofeedback" was invented, McClintock experimented with ways to control her own temperature and blood flow, until, in time, she began to feel a sense of what it took.

But these interests were not popular. "I couldn't tell other people at the time because it was against the 'scientific method.' ... We just hadn't touched on this kind of knowledge in our medical physiology, [and it is] very, very different from the knowledge we call the only way." What we label scientific knowledge is "lots of fun. You get lots of correlations, but you don't get the truth.... Things are much more marvelous than the scientific method allows us to conceive."

Our own method could tell us about some things, but not about others—for instance, she reflects, not about "the kinds of things that made it possible for me to be creative in an unknown way. *Why* do you know? Why were you so sure of something when you couldn't tell anyone else? You weren't sure in a boastful way; you were sure in what I call a completely internal way.... What you had to do was put it into their frame. Wherever it came in your frame, you had to work to put it into their frame. So you work with so-called scientific methods to put it into their frame *after* you know. Well, [the question is] *how* you know it. I had the idea that the Tibetans understood this *how* you know."

McClintock is not the only scientist who has looked to the East for correctives to the limitations of Western science. Her remarks on her relation to the phenomena she studies are especially reminiscent of the lessons many physicists have drawn from the discoveries of atomic physics. Erwin Schrödinger, for example, wrote: "... our science—Greek science—is based on objectification.... But I do believe that this is precisely the point where our present way of thinking does need to be amended, perhaps by a bit of blood-transfusion from Eastern thought."[3] Niels Bohr, the "father of quantum mechanics," was even more explicit on the subject. He wrote: "For a parallel to the lesson of atomic theory . . . [we must turn] to those kinds of epistemological problems with which already thinkers like the Buddha and Lao Tzu have been confronted, when trying to harmonize our position as spectators and actors in the great drama of existence."[4] Robert Oppenheimer held similar views: "The general notions about human understanding . . . which are illustrated by discoveries in atomic physics are not in the nature of being wholly unfamiliar, wholly unheard of, or new," he wrote. "Even in our culture they have a history, and in Buddhist and Hindu thought a more considerable and central place."[5] Indeed, as a result of a number of popular accounts published in the last decade, the correspondences between modern physics and Eastern thought have come to seem commonplace.[6] But among biologists, these interests are not common. McClintock is right to see them, and herself, as oddities. And here, as elsewhere, she takes pride in being different. She is proud to call herself a "mystic."

Above all, she is proud of her ability to draw on these other ways of knowing in her work as a scientist. It is that which, to her, makes the life of science such a deeply satisfying one—even, at times, ecstatic. "What is ecstasy? I don't understand ecstasy, but I enjoy it. When I have it. Rare ecstasy."

Somehow, she doesn't know how, she has always had an "exceedingly strong feeling" for the oneness of things: "Basically, everything is one. There is no way in which you draw a line between things. What we [normally] do is to make these subdivisions, but they're not real. Our educational system is full of subdivisions that are artificial, that shouldn't be there. I think maybe poets—although I don't read poetry—have some understanding of this." The ultimate descriptive task, for both artists and scientists, is to "ensoul" what one sees, to attribute to it the life one shares with it; one learns by identification.[7]

Much has been written on this subject, but certain remarks of Phyllis Greenacre, a psychoanalyst who has devoted a lifetime to studying the dynamics of artistic creativity, come especially close to the crux of the issue that concerns us here. For Greenacre, the necessary condition for the flowering of great talent or genius is the development in the young child of what she calls a "love affair with the world."[8] Although she believes that a special range and intensity of sensory responsiveness may be innate in the potential artist, she also thinks that, under appropriate circumstances, this special sensitivity facilitates an early relationship with nature that resembles and may in fact substitute for the intimacy of a more conventional child's personal relationships. The forms and objects of nature provide what Greenacre calls "collective alternatives," drawing the child into a "collective love affair."

Greenacre's observations are intended to describe the childhood of the young artist, but they might just as readily depict McClintock's youth. By her own account, even as a child, McClintock neither had nor felt the need of emotional intimacy in any of her personal relationships. The world of nature provided for her the "collective alternatives" of Greenacre's artists; it became the principal focus of both her intellectual and her emotional energies. From reading the text of nature, McClintock reaps the kind of understanding and fulfillment that others acquire from personal intimacy. In short, her "feeling for the organism" is the mainspring of her creativity. It both promotes and is promoted by her access to the profound connectivity of all biological forms—of the cell, of the organism, of the ecosystem.

The flip side of the coin is her conviction that, without an awareness of the oneness of things, science can give us at most only nature-in-pieces; more often it gives us only pieces of nature. In McClintock's view, too restricted a reliance on scientific methodology invariably leads us into difficulty. "We've been spoiling the environment just dreadfully and thinking we were fine, because we were using the techniques of science. Then it turns into technology, and it's slapping us back because we didn't think it through. We were making assumptions we had no right to make. From the point of view of how the whole thing actually worked, we knew how part of it worked. . . . We didn't even inquire, didn't even see how the rest was going on. All these other things were happening and we didn't see it."

She cites the tragedy of Love Canal as one example, the acidification of the Adirondacks Lakes as another. "We didn't think [things] through. . . . If you take the train up to New Haven . . . and the wind is from the southeast, you find all of the smog from New York is going right up to New Haven. . . . We're not thinking it through, just spewing it out. . . . Technology is fine, but the scientists and engineers only partially think through their problems. They solve certain aspects, but not the total, and as a consequence it is slapping us back in the face very hard."

• • •

Barbara McClintock belongs to a rare genre of scientist; on a short-term view of the mood and tenor of modern biological laboratories, hers is an endangered species. Recently, after a public seminar McClintock gave in the Biology Department at Harvard University, she met informally with a group of graduate and postdoctoral students. They were responsive to her exhortation that they "take the time and look," but they were also troubled. Where does one get the time to look and to think? They argued that the new technology of molecular biology is self-propelling. It doesn't leave time. There's always the next experiment, the next sequencing to do. The pace of current research seems to preclude such a contemplative stance. McClintock was sympathetic, but reminded them, as they talked, of the "hidden complexity" that continues to lurk in the most straightforward-seeming systems. She herself had been fortunate; she had worked with a slow technology, a slow organism. Even in the old days, corn had not been popular because one could never grow more than two crops a year. But after a while, she'd found that as slow as it was, two crops a year was too fast. If she was really to analyze all that there was to see, one crop was all she could handle.

There remain, of course, always a few biologists who are able to sustain the kind of "feeling for the organism" that was so productive—both scientifically and personally—for McClintock, but to some of them the difficulties of doing so seem to grow exponentially. One contemporary, who says of her own involvement in research, "If you want to really understand about a tumor, you've got to *be* a tumor," put it this way: "Everywhere in science the talk is of winners, patents, pressures, money, no money, the rat race, the lot; things that are so completely alien . . . that I no longer know whether I can be classified as a modern scientist or as an example of a beast on the way to extinction."[9]

McClintock takes a longer view. She is confident that nature is on the side of scientists like herself. For evidence, she points to the revolution now occurring in biology. In her view, conventional science fails to illuminate not only "how" you know, but also, and equally, "what" you know. McClintock sees additional confirmation of the need to expand our conception of science in her own—and now others'—discoveries. The "molecular" revolution in biology was a triumph of the kind of science represented by classical physics. Now, the necessary next step seems to be the reincorporation of the naturalist's approach—an approach that does not press nature with leading questions but dwells patiently in the variety and complexity of organisms. The discovery of genetic lability and flexibility forces us to recognize the magnificent integration of cellular processes—kinds of integration that are "simply incredible to our old-style thinking." As she sees it, we are in the midst of a major revolution that "will reorganize the way we look at things, the way we do research." She adds, "And I can't wait. Because I think it's going to be marvelous, simply marvelous. We're going to have a completely new realization of the relationship of things to each other."

Notes

Preface

1. Marcus Rhoades, "Barbara McClintock: Statement of Achievements," Statement for the National Academy of Sciences, 1967 (unpublished).
2. Quoted in Horace Freeland Judson, *The Eighth Day of Creation: Makers of the Revolution in Biology* (New York: Simon and Schuster, 1979), p. 216.
3. Matthew Meselson, private interview, December 18, 1979.
4. Thomas S. Kuhn, *The Structure of Scientific Revolutions* (Chicago: University of Chicago Press, 1970), p. 152.
5. Ibid., p. 153.
6. All quotations from Barbara McClintock are taken from private interviews conducted between September 24, 1978, and February 25, 1979.

Chapter 1

1. Charles Rosenberg, "Factors in the Development of Genetics in the United States: Some Suggestions," *Journal of the History of Medicine of Allied Sciences* 22 (1967):27–46.

2. Mordecai L. Gabriel and Seymour Fogel, *Great Experiments in Biology* (Englewood Cliffs, N.J.: Prentice-Hall, 1955), p. 268.
3. James A. Peters, ed., *Classic Papers in Genetics* (Englewood Cliffs, N.J.: Prentice-Hall, 1959), p. 156.
4. Gunther Stent, *The Coming of the Golden Age* (New York: American Museum of Natural History Press, 1969), p. 343.
5. Gunther Stent, "That Was the Molecular Biology That Was," *Science* 16 (April 26, 1968):393.
6. Ibid.
7. Judson, op. cit., p. 613.
8. Francis H. C. Crick, "On Protein Synthesis," *Symposium of the Society of Experimental Biology* 12 (1957):138–163.
9. Judson, op. cit., p. 461.
10. Barbara McClintock, "Some Parallels Between Gene Control Systems in Maize and in Bacteria," *American Naturalist* 95 (1961):266.
11. Peters, op. cit., p. 156.
12. Judson, op. cit., p. 461.
13. From Monod's concluding summary in Jacques Monod and François Jacob, "Teleonomic Mechanism's Cellular Metabolism, Growth, and Differentiation," *Cold Spring Harbor Symposia on Quantitative Biology* 26 (1961):394–395.
14. François Jacob and Jacques Monod, "On the Regulation of Gene Activity: β-Galactosidase Formation in *E. coli,*" *Cold Spring Harbor Symposia on Quantitative Biology* 26 (1961):207.
15. Quoted in Court Lewis, "She Asks, What Makes a Gene Work," *Carnegie Institution of Washington Newsletter* (December 1978):4–5.
16. Quoted in *Newsweek,* November 30, 1981, p. 74.

Chapter 2

1. D. W. Winnicott, "The Capacity to Be Alone," *International Journal of Psychoanalysis* 30 (1958):416–420.
2. Boston University was known as a "homeopathic" medical school, and Dr. McClintock was listed in the 1900 Polk's Medical Register as a homeopathic doctor—a more accepted category of medical practice then than now.
3. Quoted in "A Tribute to Henry Sage from the Women Graduates of Cornell, 1895" (the quote is from May 15, 1973).
4. Inscription on the cornerstone of Sage College, Cornell University.

Chapter 3

1. Marcus Rhoades, "Barbara McClintock: Statement of Achievements," Statement for the National Academy of Sciences, 1967 (unpublished).
2. Marcus Rhoades, private interview, May 16, 1980.
3. George Beadle, private interview, May 15, 1980.
4. George Beadle, "Biochemical Genetics: Some Recollections," in John Cairns, Gunther Stent, and James Watson, eds., *Phage and the Origins of Molecular Biology* (Cold Spring Harbor, N.Y.: Cold Spring Harbor Laboratory of Quantitative Biology, 1966), p. 24.
5. The academic opportunities available to women in the 1920s can be gleaned from an early report of the American Association of University Professors. In 1921, only .001 percent of the professorships at the coeducational colleges and universities were held by women. Most of the latter positions were in home economics and physical education. By contrast, women held 68 percent of the professorships at the women's colleges. Change in this profile did not begin to occur until after World War II, and then only slowly and sporadically. Margaret Rossiter, "Women Scientists in America Before 1920," *American Scientist* 62 (1974):315.
6. Harriet Creighton, private interview, January 8, 1980.
7. Rossiter, op. cit.
8. Curt Stern, "From Crossing-Over to Developmental Genetics," in Lewis Stadler, *Stadler Symposium, Vol. 1* (Columbia, Mo.: The Curator of the University of Missouri, 1971), p. 24.
9. T. H. Morgan, "Opening Address," in Donald F. Jones, ed., *Proceedings of the Sixth International Congress of Genetics, Vols. I and II* (Menosha, Wis.: Brooklyn Botanical Garden, 1932), pp. 102–103.

Chapter 4

1. Marcus Rhoades, private interview, May 16, 1980.
2. Barbara McClintock, "The Relation of a Particular Chromosomal Element to the Development of the Nucleoli in *Zea mays,*" *A. Zellforsch. u. Mikr. Anat.* 21 (1934):294–328.
3. F. G. Jordan, "The Nucleolus at Weimar," *Nature* 281 (1979):529–530.
4. Harriet Creighton, private interview, January 8, 1980.

5. Quoted in Warren Weaver's diary, April 24, 1934, Rockefeller Foundation Archives, 205D, CIT 1934, January–June, R.G. 1.1, Series 200, Folder 72.
6. Warren Weaver's diary, June 24, 1934, Rockefeller Foundation Archives, R.G. 1.1, Series 200, Box 136, Folder 1679.
7. Letter from A. R. Mann to Frank Blair Hanson, August 14, 1935, Rockefeller Foundation Archives, R. G. 1.1, Series 200, Box 136, Folder 1680.
8. C. W. Metz, May 2, 1935, Rockefeller Foundation Archives, R.G. 1.1, Series 200, Box 136, Folder 1680.
9. Frank Blair Hanson's diary, November 21, 1935, Rockefeller Foundation Archives, R.G. 1.1, Series 200, Box 136, Folder 1680.
10. Frank Blair Hanson's diary, August 7, 1935, Rockefeller Foundation Archives, R.G. 1.1, Series 200, Box 136, Folder 1680.
11. Frank Blair Hanson's notes at the Genetics Meeting in Woods Hole, Mass., August 14–27, 1935, Rockefeller Foundation Archives, R.G. 1.1, Series 200, Box 136, Folder 1680.
12. L. C. Dunn, *A Short History of Genetics: The Development of Some Main Lines of Thought, 1864–1939* (New York: McGraw-Hill, 1965), p. 164.

Chapter 5

1. In 1937–1938, McClintock's annual salary is listed in the Official Manual of the State of Missouri at $2700. No record can be found of her salary between 1935 and 1937, but in 1942, when she resigned, Stadler wrote to the Dean, saying: "It will be impossible to replace her with a person of comparable attainments at the salary which she received." Manuscript collection #2429, L. J. Stadler Papers, June 13, 1942, University of Missouri, Western Historical Manuscript Collection, Columbia State Historical Society of Missouri Manuscripts.
2. Marcus Rhoades, private interview, May 16, 1980.
3. Frank Blair Hanson's notes, July 11–August 31, 1940, Rockefeller Foundation Archives, 200D, R.G. 1.1, Series 200, Box 160, Folder 1967.
4. This recollection is at least partially confirmed in Stadler's letter to Curtis concerning her resignation and replacement. He refers to the money originally slated for McClintock as "considerably less

than her salary would have been if she had accepted the appointment recently offered her." (See note 1.)

5. Quoted in Gerald Holton, *Thematic Origins of Scientific Thought: Kepler to Einstein* (Cambridge, Mass.: Harvard University Press, 1973), p. 377.

Chapter 6

1. C. D. Darlington, "Recent Advances in Cytology," reprinted in C. D. Darlington, *Cytology,* 2nd ed. (London: J. & A. Churchill, 1965), p. 15.
2. C. D. Darlington, "Chromosomes as *We* See Them" (Opening Address), in C. D. Darlington and K. R. Lewis, *Chromosomes Today, Vol. I: Proceedings of the First Oxford Chromosome Conference, July 28–31, 1964* (New York: Plenum Press, 1966), pp. 3–4.
3. The term "organizer" was introduced by Hans Spemann and Hilda Mangold in 1924 to account for the effects of transplanting critical pieces of tissue in salamander embryos. It became a guiding concept in embryological thought for the next three decades.
4. L. C. Dunn, *A Short History of Genetics: The Development of Some Main Lines of Thought, 1864–1939* (New York: McGraw-Hill, 1965), p. 188.
5. T. H. Morgan, "Chromosomes and Heredity," *American Naturalist* 44 (1910):449–496.
6. T. H. Morgan, *Embryology and Genetics* (New York: Columbia University Press, 1934).
7. Quoted in Garland Allen, *Life Science in the Twentieth Century* (Cambridge, England: Cambridge University Press, 1975), p. 125.
8. Ross Harrison, "Embryology and Its Relation to Genetics," *Science* 85 (1937):369.
9. Ernst Mayr, "Evolution," *Scientific American* (September 1978): 52.
10. Thomas Kuhn, *The Structure of Scientific Revolutions* (Chicago: University of Chicago Press, 1962), p. 151
11. Curt Stern, "Richard Benedict Goldschmidt," *Biographical Memoirs* (Washington, D. C.: National Academy of Sciences, 1967), p. 165.
12. Garland Allen, "Opposition to the Mendelian Chromosome Theory: The Physiological and Developmental Genetics of Rich-

ard Goldschmidt," *Journal of the History of Biology* 7 (1974): 49–92.
13. Stephen Jay Gould, "The Return of Hopeful Monsters," *Natural History* 86 (1977):24.

Chapter 7

1. Vannevar Bush, "President's Report of 1942" *Carnegie Institution of Washington Yearbook* 41 (1942):3.
2. Letter from Barbara McClintock to Tracy Sonneborn, May 14, 1944.
3. Quoted in Warren Weaver's diary, April 10–20, 1946, Rockefeller Foundation Archives, Series 200, R.G. 1.1., Box 160, Folder 1968.
4. Quoted in Holton, *Thematic Origins of Scientific Thought,* op. cit., p. 378.

Chapter 8

1. Barbara McClintock, "Maize Genetics," *Carnegie Institution of Washington Yearbook* 45 (1946):180.
2. Ibid., p. 182.
3. Ibid., p. 186.
4. Barbara McClintock, "Cytogenetic Studies of Maize and *Neurospora,*" *Carnegie Institution of Washington Yearbook* 46 (1947): 147.
5. Ibid.
6. Barbara McClintock, "Mutable Loci in Maize," *Carnegie Institution of Washington Yearbook* 47 (1948):159
7. Ibid., p. 160.
8. Ibid., p. 159.
9. Barbara McClintock, "Mutable Loci in Maize," *Carnegie Institution of Washington Yearbook* 48 (1949):142–143.
10. Ibid., p. 143.
11. Barbara McClintock, "Chromosome Organization and Genic Expression," *Cold Spring Harbor Symposia on Quantitative Biology* 16 (1951):40.
12. Ibid.
13. Ibid., p. 42.
14. Ibid., p. 34.

15. Evelyn Witkin, private interview, April 14, 1980.

Chapter 9

1. Barbara McClintock, "Induction of Instability of Selected Loci in Maize," *Genetics* 38 (1953):579–599.
2. Royal A. Brink and Robert A. Nilan, "The Relation Between Light Variegated and Medium Variegated Pericarp in Maize," *Genetics* 37 (1952):519–544; P. C. Barclay and R. A. Brink, "The Relation Between Modulator and Activator in Maize," *Proceedings of the National Academy of Sciences* 40 (1954):1118–1126; Peter Peterson, "A Mutable Pale Green Locus in Maize," *Genetics* 38 (1953): 682–683; and Peter Peterson, "The Pale Green Mutable System in Maize," *Genetics* 45 (1960):115–132.
3. Lotte Auerbach, private interview, April 10, 1981.
4. Quoted in Holton, *Thematic Origins of Scientific Thought,* op. cit., p. 357.
5. Some of these issues are eloquently discussed by Michael Polanyi in his essay, "The Unaccountable Element in Science," in Marjorie Green, ed., *Knowing and Being* (Chicago: University of Chicago Press, 1969), pp. 123–137.
6. Freeman Dyson, *Disturbing the Universe* (New York: Harper & Row, 1980), p. 54.
7. Ibid., p. 55.
8. Ibid., pp. 55–56.
9. Rudolf Arnheim, *Art and Visual Perception* (Berkeley: University of California Press, 1954), p. 442.
10. Gerald Holton, *The Scientific Imagination: Case Studies* (London: Cambridge University Press, 1974), p. 38.
11. Ibid.
12. Quoted in Holton, *Thematic Origins of Scientific Thought,* op. cit., p. 358.
13. Ibid., p. 368.
14. Ibid.

Chapter 10

1. Milislav Demerec, "Foreword," *Cold Spring Harbor Symposia on Quantitative Biology* 16 (1951):v.
2. Ibid.

3. Richard Goldschmidt, "Chromosomes and Genes," *Cold Spring Harbor Symposia on Quantitative Biology* 16 (1951):1.
4. Barbara McClintock, "The Origin and Behavior of Mutable Loci in Maize," *Proceedings of the National Academy of Sciences* 36 (1950):347.
5. Goldschmidt, op. cit., p. 3.
6. Ibid., p. 4.
7. Lewis Stadler, "The Gene," *Science* 120 (1954):811–819.
8. Ibid., p. 814.
9. Ibid., p. 818.
10. Demerec, op. cit., p. v.
11. James D. Watson, *Molecular Biology of the Gene,* 2nd ed. (New York: W. A. Benjamin, 1970).
12. Emory L. Ellis, "Bacteriophage: One-Step Growth," in Cairns, Stent, and Watson, op. cit., p. 58.
13. Emory L. Ellis and Max Delbrück, "The Growth of Bacteriophage," *Journal of General Physiology* 22 (1939):365.
14. Max Delbrück, "Experiments with Bacterial Viruses (Bacteriophages)," *Harvey Lectures* 41 (1946):161–162.
15. Ibid.
16. Gunther Stent, *Molecular Biology of Bacterial Viruses* (San Francisco: W. H. Freeman and Company, 1963), p. 376.
17. Stent, "That Was the Molecular Biology That Was," op. cit., p. 393.
18. Ibid., p. 363.
19. Aaron Novick, "Phenotypic Mixing," in Cairns, Stent, and Watson op. cit., pp. 134–135.
20. Erwin Schrödinger, *What Is Life? & Mind and Matter* (London: Cambridge University Press, 1945).
21. James D. Watson, "Growing Up in the Phage Group," in Cairns, Stent, and Watson, op. cit., p. 240.
22. Ibid., p. 241.
23. Stent, "That Was the Molecular Biology That Was," op. cit, p. 393.
24. Max Delbrück, "A Physicist Looks at Biology," in Cairns, Stent, and Watson, op. cit., p. 22.
25. Ibid.
26. Ibid.
27. Ibid.
28. James D. Watson and Francis H. C. Crick, "Genetical Implications of the Structure of Deoxyribonucleic Acid," *Nature* 171 (1953): 964–967.

29. Francis H. C. Crick, "On Protein Synthesis," *Symposium of the Society of Experimental Biology* 12 (1957):153.
30. Quoted in Judson, op. cit., p. 217.

Chapter 11

1. Watson and Crick, op. cit.
2. *Life* Magazine, May 1980.
3. B. McClintock, "Controlling Elements and the Gene," in *Cold Spring Harbor Symposia on Quantitative Biology* 21 (1956): 215.
4. Quoted in Judson, op. cit., p. 353.
5. François Jacob and Jacques Monod, "Genetic Regulatory Mechanisms in the Synthesis of Proteins," *Journal of Molecular Biology* 3 (1961):356.
6. Barbara McClintock, "Some Parallels Between Gene Control Systems in Maize and in Bacteria," *American Naturalist* 95 (1961): 276.
7. Quoted in Fred Wilcox, "Everyone Suddenly Pays Homage to a Geneticist Most Persistent," *Cornell Alumni News* 84 (February 1982):4.
8. A. L. Taylor, "Bacteriophage-Induced Mutation in *E. coli*," *Proceedings of the National Academy of Sciences* 50 (1963):1043.
9. Jonathan Beckwith, Ethan Signer, and Wolfgang Epstein, "Transposition of the *lac* Region of *E. coli*," *Cold Spring Harbor Symposia on Quantitative Biology* 31 (1966):393.
10. James A. Shapiro, S. L. Adhya, and Ahmad I. Bukhari, "New Pathway in the Evolution of Chromosome Structure," in Ahmad I. Bukhari, James A. Shapiro, and S. L. Adhya, eds., *DNA Insertion Elements, Plasmids and Episomes* (Cold Spring Harbor, N.Y.: Cold Spring Harbor Laboratory of Quantitative Biology, 1977).
11. Allan Campbell et al., "Nomenclature of Transposable Elements in Prokaryotes," in Bukhari et al., op. cit., p. 16.
12. Heinz Saedler, "IS1 and IS2 in *E. coli:* Implications for the Evolution of the Chromosome and Some Plasmids," in Bukhari et al., op. cit., pp. 65–72.
13. Patricia Nevers and Heinz Saedler, "Transposable Genetic Elements as Agents of Gene Instability and Chromosomal Rearrangements," *Nature* 268 (1977):109.

14. One such study is Gerald Fink et al., "Transposable Elements in Yeast," *Cold Spring Harbor Symposia on Quantitative Biology* 45 (1981):575–580.
15. James A. Shapiro, "Changes in Gene Order and Gene Expression," paper presented at the First International Symposium on Research Frontiers in Aging and Cancer, Washington, D.C., September 21–25, 1980, p. 29.
16. James A. Shapiro, "Reflections on the Information Content of Chromosome Structure and How It Changes" (in press).
17. Barbara McClintock, "Mechanisms That Rapidly Reorganize the Genome," in G. P. Reder, ed., *Stadler Symposium, Vol. 10* Columbia, Mo.: The Curator of the University of Missouri, 1978), pp. 25–48.
18. McClintock, "Mechanisms That Rapidly Reorganize the Genome," op. cit., p. 26.
19. Barbara McClintock, "Modified Gene Expressions Induced by Transposable Elements," in W. A. Scott et al., eds., *Mobilization and Reassembly of Genetic Information* (New York: Academic Press, 1980), pp. 11–19, p. 17.
20. Quoted in Jean L. Marx, "A Movable Feast in the Eukaryotic Genome," *Science* 211 (1981):153.

Chapter 12

1. Quoted in E. Broda, "Boltzman, Einstein, Natural Law and Evolution," *Comparative Biochemical Physiology* 67B (1980):376.
2. Quoted in Banesh Hoffmann and Helen Dukas, *Albert Einstein, Creator and Rebel* (New York: New American Library, 1973), p. 222.
3. Schrödinger, *What Is Life?*, op. cit., p. 140.
4. Niels Bohr, *Atomic Physics and Human Knowledge* (New York: John Wiley and Sons, 1958), p. 33.
5. Robert J. Oppenheimer, *Science and the Common Understanding,* (New York: Simon and Schuster, 1954), pp. 8–9.
6. See, for example, Fritjof Capra, *The Tao of Physics* (Berkeley, Ca.: Shambhala, 1975), and Gary Zukov, *The Dancing Wu Li Masters* (New York: William Morrow, 1979).
7. The word "ensoul" is taken from Marion Milner, who wrote of her own endeavors as an artist: "I want to ensoul nature with what was

was really there." Marion Milner, *On Not Being Able to Paint* (New York: International Universities Press, 1957), p. 120.

8. Phyllis Greenacre, "The Childhood of the Artist: Libidinal Phase Development and Giftedness" (1957), reprinted in Phyllis Greenacre, *Emotional Growth: Psychoanalytic Studies of the Gifted and a Great Variety of Other Individuals* (New York: International Universities Press, 1971), p. 490.
9. June Goodfield, *An Imagined World: A Story of Scientific Discovery* (New York: Harper & Row, 1981), p. 213.

Glossary

Acentric fragments Fragments of chromosomes without centromeres.

Allele One of two (or more) alternate forms of a gene.

Anaphase The stage of mitosis or meiosis when daughter chromosomes separate and begin to move toward the poles of the spindles.

Asci In *Ascomyceta* fungi, the saclike cells in which fusion of haploid nuclei occur during sexual reproduction, followed by meiosis and formation of haploid spores.

Bacteriophages (phages) Viruses that multiply in bacteria.

Biometrics The application of statistics to biological problems.

Centromere The region of the chromosome that is attached to the spindle during mitosis or meiosis; a thin, dense, platelike body.

Chiasmata (cross points) Visible points of interchange between chromatids of homologous chromosomes during prophase in meiosis. See *Crossing over.*

Chromatid One of the two strands that result from the duplication of a chromosome during the prophase of nuclear division.

Chromomeres Darkly staining granules found at consistent locations along a chromosome. They are useful cytological markers for mapping gene location on the chromosome.

Chromosome A threadlike body in the nucleus of cells that can be seen at cell division. Chromosomes carry the genes.

Cistron The shortest length of genetic material (DNA nucleotide sequence) that comprises a functional unit.

Crossing over In meiosis, each chromosome lines up with its complementary chromosome (conjugation), then chromosomes partially separate (diplotene). Exchange of chromatids, or crossing over, occurs at cross points, or chiasmata, where the chromosomes can be seen to cross.

Cytogenetics The science that links the study of the visible structure of the chromosomes with genetics.

Cytology The microscopic study of cell structure.

Deletion (deficiency) The loss of part of a chromosome.

Determination events A change occurring in a cell that would show up in the progeny of that cell many generations later. An irreversible event responsible for differentiation of the cell later on.

Diakinesis The final stage in prophase of first division of meiosis following diplotene.

Dicentric Designating a chromosome or chromatid having two centromeres.

Differentiation The process of change in cells during development, resulting in structures and functions that characterize the different kinds of cells in different parts of an organism of some later phase of life history.

Diploid Having the chromosomes in pairs, the members of each pair being homologous, so that twice the haploid number is present.

Diplotene The stage in prophase of first division of meiosis following pachytene, in which pairs of chromatids derived from homologous chromsomes begin to separate from each other except at certain points of connection (chiasmata) where interchange occurs between chromatid segments (crossing over).

Dissociation A regular break in the chromosome occurring at a highly specific location.

DNA (deoxyribonucleic acid) The molecular basis of heredity. A chain of deoxyribonucleotides, which, according to the Watson-Crick model, forms a double helix held together by hydrogen bonds between specific pairs of bases. Genetic information is contained in the sequence of bases. Each strand in the double helix is complementary to its partner strand in terms of its base sequence.

Drosophila Fruit fly. Widely used in the study of inheritance because of its short generation span.

Endosperm Nutritive tissue surrounding and nourishing the embryo in seed plants.

Enzyme A protein that acts as a catalyst.

Eukaryotes Cells or organisms in which the nucleus is separated from the rest of the cell by a nuclear membrane.

Gamete A reproductive cell (usually haploid) whose nucleus and cytoplasm fuse with those of another gamete during fertilization. The resulting cell (zygote) develops into a new organism.

Gametophyte A phase in the life cycle of plants having haploid nuclei, during which the sex cells are produced.

Gene (hereditary factor) The unit of inheritance.

Gene locus (gene site) The position that a gene occupies on a chromosome.

Genetic code (chemical code) The system of correspondence between nucleotide sequence and amino-acid sequence. Each of the twenty amino acids (the building blocks of protein) is specified by a sequence of three adjacent nucleotides.

Genome (chromosomal complement) The set of chromosomes found in each nucleus in the cells of an organism of a given species.

Haploid Having a single set of unpaired chromsomes in each nucleus, characteristic of gametes.

Heterozygous Having two different alleles at the two corresponding loci in a pair of homologous chromosomes.

Homologous chromosomes Chromosomes containing identical sets of loci. Two homologous chromosomes or parts of chromosomes have a strong attraction for each other during early stages of meiosis and undergo pairing.

Homozygous Having identical alleles at the two corresponding loci of homologous chromosomes.

Hybrid A plant or an animal resulting from a cross between parents that are genetically unlike, often the offspring of two different species or well-marked varieties within a species.

Interphase The state of a cell when it is not undergoing mitosis.

Inversion Reversal of part of a chromosome so that genes within that part lie in inverse order.

Lamarckism The view that acquired characters (characteristics that appeared in the parents as a result of environmental influences) are inherited by the offspring and that the cumulative effect of the inheritance of these characters is the mechanism of evolutionary change.

Life cycle The series of developmental changes undergone by an organism from fertilization to reproduction and death.

Linkage The greater association in inheritance of two or more nonallelic genes than is to be expected from independent assortment. Genes are linked because they reside on the same chromosome.

Linkage group All the genes on one chromosome.

Linkage map (chromosome map) A plan showing the position of the genes on a chromosome.

Lysenkoism The doctrine promulgated by the Soviet scientist Trofin Lysenko between 1932 and 1965. Lysenko did not accept the gene concept and argued instead for the inheritance of characteristics acquired during the lifetime of the organism.

Lysis A state occurring when a host cell bursts after it has been invaded by a virus, releasing the viruses that have been manufactured in the infected cell.

Meiosis The process of two successive cell divisions involving the reduction of the number of chromosomes from the diploid number to the haploid number.

Mendelism The science of the behavior of genes in inheritance studied by breeding experiments and governed by the laws originally formulated by Gregor Mendel.

Metaphase A stage of mitosis and meiosis when chromosomes are arranged on the equator of the spindle.

Microtubules Fibrous structures contained in the spindles, responsible for the separation of chromatids during anaphase.

Mitosis The usual process by which cell nuclei divide into two with the formation of daughter nuclei having an identical chromosomal complement to that of the original nucleus.

Molecular biology A modern branch of biology concerned with explaining biological phenomena in molecular terms.

Mosaicism A phenomenon exhibited in organisms whose tissues are of two or more genetically different kinds. Mosaicism can occur as a result of mutation or abnormal distribution of chromosomes, affecting one cell and all its descendents during development.

Mutation An inherited change in a gene or chromosome set.

Nucleolus A small, dense body residing in the nucleous containing RNA and protein, concerned with the synthesis of ribosomes.

Nucleoprotein The main chemical constituent of the nucleus, a combination of nucleic acid and protein.

Nucleotides The building blocks of DNA and RNA. The arrangement of the nucleotide bases contains the code for the structure of proteins. See *Genetic code.*

Nucleus The spheroidal structure present in most cells that contains the chromosomes.

Operator gene One of the genes of an operon system having the specialized function of turning the structural gene on or off. See *Operon.*

Operon A group of closely linked genes responsible for the synthesis of a group of enzymes functionally related as members of one enzyme system.

Pachytene The state in prophase of first division of meiosis following zygotene, in which each of the paired chromosomes in synapsis begins to shorten and thicken and now appears doubled as two chromatids.

Perithecium (fruiting body) Rounded or flask-shaped fruitbody of certain fungi and lichen containing the asci, which produce spores.

Phenotype The physical manifestation of a genetic trait.

Plasmid An extra chromosomal fragment containing genetic material found in certain bacteria.

Prokaryotes Unicellular organisms having no nuclear membrane separating the primary genetic material from the cytoplasm. Bacteria and blue-green algae are prokaryotes.

Prophase The initial stage of mitosis or meiosis, during which chromosomes appear within the nucleus; in meiosis they undergo pairing.

Recombination The formation of a new association of genes (e.g., by crossing over), different from that present in either parent.

Regulator gene A gene lying outside the operon system, elsewhere on the chromosome, that codes for the repressor substance (a protein). The repressor inhibits the operator gene from "turning on" the structural gene.

Repressor A protein, produced by a regulator gene, that inhibits the action of the operator gene.

Ribosomes The "factories" of protein synthesis within the cell. Messenger RNA attaches to the ribosomes and there receives molecules of transfer RNA bearing amino acids, which are then assembled into proteins.

Ring chromosome A chromosome fragment that has fused (annealled) its own ends together, forming a ring.

RNA (ribonucleic acid) A chain of ribonucleotides. RNA molecules are single stranded and are found in three classes: (1) messenger RNA (complementary to the DNA of structural genes), (2) ribosomal RNA, and (3) transfer RNA (the molecule that transfers an amino acid to the protein chain specified by the messenger RNA).

Semiconservative replication The method of replication of DNA in which the molecule divides longitudinally, each half being conserved and acting as a template for the formation of a new strand.

Somatic (cells) The diploid body cells of an organism, as distinct from the germ cells.

Spindle A structure mediating the separation of chromatids during mitosis or meiosis.

Structural gene The gene that codes for the synthesis of a protein.

Synapsis (conjugation) Side-by-side association of homologous chromosomes at meiosis. See *Diplotene.*

Systematics The classification of organisms based on their evolutionary relationships.

Telophase The terminal stage of mitosis or meiosis during which nuclei revert to the resting stage (interphase).

Transduction The process that occurs when bacteriophage carry pieces of genetic material from one bacterial chromosome to another.

Translocation Transfer of a part of a chromosome into (usually) a nonhomologous chromosome.

Variegation Irregular variation of color in leaves and flowers or kernels, due to suppression of normal pigment production.

Viruses Infectious disease-causing agents, smaller than bacteria, that always require intact host cells for replication and that contain either DNA or RNA as their genetic component.

Zygote The diploid cell resulting from the fusion of two haploid gametes (fertilization), before it undergoes any further differentiation, that is, a fertilized egg cell.

Zygotene The stage in prophase of first division of meiosis following leptotene, in which pairing (synopsis) of homologous chromosomes occurs.

Index